Statistical Modeling and Applications on Real-Time Problems

In an era dominated by mathematical and statistical models, this book unravels the profound significance of these tools in decoding uncertainties within numerical, observational, and calculation-based data. From governmental institutions to private entities, statistical prediction models provide a critical framework for optimal decision-making, offering nuanced insights into diverse realms, from climate to production and beyond.

This book

- Serves as a comprehensive resource in statistical modeling, methodologies, and optimization techniques across various domains.
- Features contributions from global authors; the compilation comprises ten insightful chapters, each addressing critical aspects of estimation and optimization through statistical modeling.
- Covers a spectrum of topics, from non-parametric goodness-of-fit statistics to Bayesian applications; the book explores novel resampling methods, advanced measures for empirical mode, and transient behavior analysis in queueing systems.
- Includes asymptotic properties of goodness-of-fit statistics, practical applications of Bayesian Statistics, modifications to the Hard EM algorithm, and explicit transient probabilities.
- Culminates with an exploration of an inventory model for perishable items, integrating preservation technology and learning effects to determine the economic order quantity.

This book stands as a testament to global collaboration, offering a rich tapestry of commendable statistical and mathematical modeling alongside real-world problem-solving. It is poised to ignite further exploration, discussion, and innovation in the realms of statistical modeling and optimization.

Mathematical Engineering, Manufacturing, and Management Sciences

Series Editor: *Mangey Ram, Professor, Assistant Dean (International Affairs), Department of Mathematics, Graphic Era University, Dehradun, India*

The aim of this new book series is to publish the research studies and articles that bring up the latest development and research applied to mathematics and its applications in the manufacturing and management sciences areas. Mathematical tools and techniques are the strength of engineering sciences. They form the common foundation of all novel disciplines as engineering evolves and develops. The series will include a comprehensive range of applied mathematics and its application in engineering areas such as optimization techniques, mathematical modelling and simulation, stochastic processes and systems engineering, safety-critical system performance, system safety, system security, high assurance software architecture and design, mathematical modelling in environmental safety sciences, finite element methods, differential equations, and reliability engineering.

Biodegradable Composites for Packaging Applications
Edited by Arbind Prasad, Ashwani Kumar and Kishor Kumar Gajrani

Computing and Stimulation for Engineers
Edited by Ziya Uddin, Mukesh Kumar Awasthi, Rishi Asthana and Mangey Ram

Advanced Manufacturing Processes
Edited by Yashvir Singh, Nishant K. Singh and Mangey Ram

Additive Manufacturing: Advanced Materials and Design Techniques
Pulak M. Pandey, Nishant K. Singh and Yashvir Singh

Advances in Mathematical and Computational Modeling of Engineering Systems
Mukesh Kumar Awasthi, Maitri Verma and Mangey Ram

Biowaste and Biomass in Biofuel Applications
Edited by Yashvir Singh, Vladimir Strezov, and Prateek Negi

Lean Manufacturing and Service: Fundamentals, Applications, and Case Studies
Kanchan Das and Miranda Dixon

For more information about this series, please visit: www.routledge.com/ Mathematical-Engineering-Manufacturing-and-Management-Sciences/book-series/CRCMEMMS

Statistical Modeling and Applications on Real-Time Problems

Unraveling Insights through Advanced Analytical Techniques

Edited by
Chandra Shekhar
Raghaw Raman Sinha

CRC Press
Taylor & Francis Group
Boca Raton London New York

CRC Press is an imprint of the
Taylor & Francis Group, an **informa** business

Front cover image: BetsyChe/Shutterstock

MATLAB® and Simulink® are trademarks of The MathWorks, Inc. and are used with permission. The MathWorks does not warrant the accuracy of the text or exercises in this book. This book's use or discussion of MATLAB® or Simulink® software or related products does not constitute endorsement or sponsorship by The MathWorks of a particular pedagogical approach or particular use of the MATLAB® and Simulink® software.

First edition published 2024
by CRC Press
2385 NW Executive Center Drive, Suite 320, Boca Raton FL 33431

and by CRC Press
4 Park Square, Milton Park, Abingdon, Oxon, OX14 4RN

CRC Press is an imprint of Taylor & Francis Group, LLC

ISBN: 978-1-032-39278-3 (hbk)
ISBN: 978-1-032-41182-8 (pbk)
ISBN: 978-1-003-35665-3 (ebk)

DOI: 10.1201/9781003356653

Typeset in Sabon
by MPS Limited, Dehradun

Contents

Preface

In an age where mathematical and statistical models are indispensable for addressing societal challenges, the imperative to analyze uncertainties in numerical, observational, and calculation-based data has grown exponentially. These models, extensively utilized by governmental and private entities, provide a crucial framework for optimal decision-making. Statistical prediction models, specifically, offer insights into probabilistic future behaviors, influencing decisions in diverse realms, from climate to production, and beyond.

Against the backdrop of real-time challenges, our book, *Statistical Modeling and Applications on Real-Time Problems: Unraveling Insights through Advanced Analytical Techniques*, emerges as a comprehensive resource. It aims to enlighten researchers about recent developments in statistical modeling, methodologies, and optimization techniques across various domains.

This compilation comprises 10 insightful chapters contributed by authors globally, each delving into critical aspects of estimation and optimization using statistical modeling and methods to address real-world issues. The book covers a range of topics, from non-parametric goodness-of-fit statistics to Bayesian applications in cosmology, novel resampling methods, advanced measures for empirical mode, transient behavior analysis in queueing systems, process fairness, modifications to the Hard EM algorithm, explicit transient probabilities, public debt instruments' effects on spending in India, and an innovative perishable inventory model.

The asymptotic properties of the proposed goodness-of-fit statistic for non-parametric multiple measurement error models are derived and studied in Chapter 1. Chapter 2 presents a practical application of Bayesian Statistics in cosmology science, obtaining constraints on the free parameters of a cosmological model in light of observational datasets. Chapter 3 introduces Improved Sufficient Bootstrapping (ISB), a novel resampling method demonstrating improvement in the estimation of parameters such as Mean, Variance, Standard Deviation, and Coefficient of Variation. In Chapter 4, bias and mean squared error expressions for the ratio-type estimator and the minimum variance expression for the regression-type estimator are derived for a new measure of the empirical mode of a study variable.

Chapter 5 analyzes the transient behavior of the distribution of busy periods for a single-server queue with balking, catastrophes, and a repairable server using the properties of Bessel functions. Process fairness is addressed based on the model's reliance in Chapter 6, introducing FixOut, a human-centered and model-agnostic framework. Chapter 7 proposes modifications to the Hard EM algorithm for building the Gaussian Mixture Model, addressing the problem of convergence to a local optimum.

In Chapter 8, explicit transient probabilities of system size for a single-server queueing system with differentiated vacations and customers' impatience are obtained in terms of the modified Bessel function of the first kind using probability generating function technique, Laplace transforms, and continued fractions, incorporating properties of the confluent hypergeometric function. Chapter 9 utilizes the error correction model in ARDL form to test the effect of various public debt instruments on public spending in India from 1985 to 2018. Finally, Chapter 10 introduces an inventory model for perishable items with preservation technology and learning effect to determine the economic order quantity.

This collection serves as a testament to the collaborative efforts of contributors worldwide, offering readers a rich tapestry of commendable statistical and mathematical modeling, alongside the analysis of real-world problems and interdisciplinary studies in applied mathematics. It is our hope that *Statistical Modeling and Applications on Real-Time Problems: Unraveling Insights through Advanced Analytical Techniques* sparks further exploration, discussion, and innovation in the realms of statistical modeling and optimization.

Chandra Shekhar
Pilani, India
Raghaw Raman Sinha
Jalandhar, India

Acknowledgments

ॐ

आचार्यात् पादमादत्ते पादम् शिष्यः स्वमेधया।
कालेन पादमादत्ते पादम् सब्रह्मचारिभिः।।

This book, "Statistical Modeling and Applications on Real-Time Problems: Unraveling Insights through Advanced Analytical Techniques," is edited with the generous support of esteemed institutions, BITS Pilani and NIT Jalandhar. We express our heartfelt gratitude for the administrative and academic support provided by these institutions.

Special appreciations are extended to Prof. Mangey Ram, Series Editor; Mr. Gauravjeet Singh, Senior Commissioning Editor, CRC Press; Isha Ahuja, Editorial Assistant-Engineering, CRC Press; and Ms. Mehnaz Hussain, Editorial Assistant-Engineering, CRC Press, for their unwavering support and valuable suggestions. We also extend our sincere thanks to CRC Press for their support throughout the publication process.

Our deepest appreciation goes to the erudite reviewers whose constructive remarks have significantly enhanced the quality of the chapters. The contributors, hailing from different parts of the world, have shared diverse manuscripts related to statistical modeling, statistical methods, optimization techniques, and problems in both statistical and optimization fields. Special thanks to the contributors for sharing content-rich thoughts.

Our heartfelt thanks go to all friends, guardians, and society members who have illuminated our path to progress. To our known and unknown supporters, we express deep gratitude for their encouragement and contributions.

We extend our deepest gratitude to our parents for their inspiration, encouragement, and cooperation. Prof. Shekhar acknowledges the support of his wife, Ms. Shikha Gupta, and children, Pranjal and Tejsav, for their understanding during busy times and continuous motivation throughout the book's preparation. Dr. Sinha is grateful to his wife, Ms. Meenakshi, and son, Arav, for their love, understanding, prayers, and constant support in completing this work.

Chandra Shekhar
Raghaw Raman Sinha

Contributors

Khursheed Alam
The A.H. Siddiqi Centre for
  Advanced Research in Applied
  Mathematics and Physics
Sharda University
Greater Noida, India

Rifaqat Ali
Department of Mathematics,
  College of Science and Arts
Muhayil, King Khalid University
Abha, Saudi Arabia

Guilherme Alves
Université de Lorraine, CNRS, Inria
  Nancy G.E., LORIA
Vandoeuvre-lès-Nancy, France

Sherif I. Ammar
Department of Mathematics,
  Faculty of Science
Menofia University
Shebin El Kom, Egypt

Fabien Bernier
Université de Lorraine, CNRS, Inria
  Nancy G.E., LORIA
Vandoeuvre-lès-Nancy, France

Vaishnavi Bhargava
Université de Lorraine, CNRS, Inria
  Nancy G.E., LORIA
Vandoeuvre-lès-Nancy, France

Chitradipa Chakraborty
Department of Mathematics and
  Statistics
Indian Institute of Technology
  Kanpur
Kanpur, India

Steven Chavez
Department of Mathematics
Texas A&M University-Kingsville
Kingsville, TX, USA

Miguel Couceiro
Université de Lorraine, CNRS,
  Inria Nancy G.E., LORIA
Vandoeuvre-lès-Nancy, France

Subhra Sankar Dhar
Department of Mathematics and
  Statistics
Indian Institute of Technology
  Kanpur
Kanpur, India

Samir Ul Hassan
Assistant Professor, ASMSOC
NMIMS University
Mumbai

Christian Heumanna
Department of Statistics
Ludwig-Maximilians-Universität
  München
Ludwigstr., München, Germany

Mahesh Kumar Jayaswal
Department of Mathematics
  and Statistics
Banasthali Vidyapith
Banasthali, Rajasthan, India

Prashant Jha
National Institute of Technology
Sikkim

K. Kalidass
Department of Mathematics
Karpagam Academy of Higher
  Education
Coimbatore, Tamil Nadu, India

Rakesh Kumar
Department of Mathematics and
  Statistics
Namibia University of Science and
  Technology
Windhoek-Namibia
School of Mathematics
Shri Mata Vaishno Devi University
Katra, J & K, India

Santosh Kumar
Department of Mathematics School
  of Basic Sciences and Research
Sharda University
Greater Noida, UP, India

Suresh Kumar
Department of Mathematics
Indira Gandhi University
Meerpur, Haryana, India

Jicheng Liu
School of Mathematics and
  Statistics
Huazhong University of Science
  and Technology
Wuhan, Hubei, China

Biswambhar Mishra
Department of Economics
North Eastern Hill University
Shillong, Meghalaya,

Amedeo Napoli
Université de Lorraine, CNRS,
  Inria Nancy G.E., LORIA
Vandoeuvre-lès-Nancy, France

Rafael C. Nunes
Instituto de Física
Universidade Federal do Rio
  Grande do Sul
Porto Alegre RS, Brazil
Divisão de Astrofísica
Instituto Nacional de Pesquisas
  Espaciais
Avenida dos Astronautas, São José
  dos Campos, SP, Brazil

Hamed Olalekan
Department of Mathematics
Texas A&M University-Kingsville
Kingsville, TX, USA

Samyajoy Pal
Department of Statistics
Ludwig-Maximilians-Universität
  München
Ludwigstr., München, Germany

M. I. G. Suranga Sampath
Department of Mathematical
  Sciences, Faculty of Applied
  Sciences
Wayamba University of Sri Lanka
Kuliyapitiya, Sri Lanka

Stephen A. Sedory
Department of Mathematics
Texas A&M University-Kingsville
Kingsville, TX, USA

Chandra Shekhar
Department of Mathematics
Birla Institute of Technology and
  Science, Pilani
Rajasthan, India

Shalabh
Department of Mathematics
and Statistics
Indian Institute of Technology
Kanpur
Kanpur, India

Shivani Sharma
Department of Mathematics
Netaji Subhas University of
Technology
Dwarka, New Delhi, India

Sarjinder Singh
Department of Mathematics
Texas A&M University-Kingsville
Kingsville, TX, USA

Bhupender Kumar Som
Jagan Institute of Management
Rohini, New Delhi, India

About the editors

Chandra Shekhar is a distinguished academician and the former Head of the Department of Mathematics at BITS Pilani, India. He is actively engaged in both research and teaching, and his expertise encompasses a wide range of mathematical fields. These areas include Queueing Theory, Computer and Communication Systems, Machine Repair Problems, Reliability and Maintainability, Stochastic Processes, Evolutionary Computation, Statistical Analysis, and Fuzzy Set & Logic. At both the undergraduate and postgraduate levels, he imparts knowledge in subjects such as Probability and Statistics, Differential Equations, Linear Algebra, Advanced Calculus, Complex Variables, Fuzzy Logic, Operations Research, Statistical Inference, and more. He is a pioneer in Evolutionary Computation, Markovian and Stochastic Modeling, Queuing Analysis and its Applications, Inventory Theory, and Reliability Theory. He actively participates in national and international conferences, Faculty Development Programs (FDPs), and has taken the lead in organizing numerous conferences, workshops, and symposiums as a convener and organizing secretary. He has received the Best Research Paper Award at an international conference. His contributions extend to scholarly publications, with over 50 research articles published in esteemed journals within these domains. Additionally, he has supervised three Ph.D. theses and authored book chapters in edited books published by globally recognized publishers. Chandra Shekhar's authorship includes textbooks such as *Differential Equations, Calculus of Variations, and Special Functions*, as well as edited books titled *Mathematical Modeling and Computation of Real-Time Problems: An Interdisciplinary Approach* and *Modeling and Applications in Operations Research*. Beyond his academic roles, Chandra Shekhar actively serves as a member of editorial boards and as a reviewer for prestigious journals and academic societies. He also contributes his expertise to various academic committees, including Hon'ble Governor/Chancellor nominees, the Board of Management, Doctoral Research Committees, Board of Studies, advisory boards, faculty selection committees, and examination boards for government and private universities, institutions, and research laboratories. As a professional, he has collaborated with and visited several renowned organizations, including IIRS (ISRO), CSIR-IIP, NIH, WIHG, CPWD, NTPC, Bank of Maharashtra, and APS Lifetech.

R. R. Sinha received his doctorate in "Sampling Techniques" from the Department of Statistics at Banaras Hindu University, Varanasi, India in 2001. He is currently an Associate Professor in the Department of Mathematics at Dr. B. R. Ambedkar National Institute of Technology, Jalandhar, India. In addition to overseeing the dissertations of M.Tech. (Artificial Intelligence) and M.Sc. (Mathematics) candidates, he has guided two Ph.D. and three M.Phil. candidates. Dr. Sinha has more than 31 research publications and three book chapters published on various topics related to sampling techniques in national and international journals and proceedings. He has presented more than 25 research papers in international seminars/conferences/symposia and has been awarded the first prize for the best oral presentation of a research paper at an international conference. He is currently a life member of the Indian Society for Probability and Statistics, Indian Society of Agricultural Statistics, Indian Statistical Association, International Indian Statistical Association, and Ramanujan Mathematical Society. In addition to his scholarly responsibilities, Dr Sinha is an active member of editorial boards and a reviewer for esteemed journals and academic associations. His areas of specialization are Sampling Theory, Data Analysis and Inference. ORCID identifier number of Dr. R. R. Sinha is 0000-0001-6386-1973.

Chapter 1

Goodness of fit based and variable selection in non-parametric measurement error model

Shalabh, Subhra Sankar Dhar, Chitradipa Chakraborty, and Prashant Jha

1.1 INTRODUCTION

Most statistical analyses assume the availability of correct observations on the random variables. In practice, this assumption is more often violated, and it is practically not possible to collect the correct or true values of data on variables, which may happen due to several reasons. One possible reason is the presence of measurement errors in the data. The difference between the true and observed values of a variable is termed as measurement error or errors-in-variables. There are several reasons which make the presence of measurement errors a rule rather than an exception. For example, sometimes the random variable is conceptually well defined, but it is hard to obtain correct observations on it. The data on the variable may be observable in categories or some proxy variable is used in place of the original variable. The presence of measurement errors in the data disturbs the properties of the statistical procedures. For instance, in the context of linear models, the ordinary least squares estimator (or equivalently the maximum likelihood estimator under normally distributed random errors) becomes a biased and inconsistent estimator of regression coefficient, whereas it is the best linear unbiased estimator of regression coefficient in the absence of measurement errors in the data.

Various parametric and non-parametric procedures have been utilized in the literature to obtain consistent estimators of the involved parameters in measurement error models. Parametric procedures based on linear regression models have been extensively explored in the literature. In the context of the linear regression model, it is not possible to consistently estimate the regression parameters without knowing some external information like the variance of the measurement error associated with independent or dependent variables, reliability ratio associated with the explanatory variables, etc., see [1]. The reliability ratio associated with the explanatory variables is defined as the ratio of true and observed variances of explanatory variables.

DOI: 10.1201/9781003356653-1

1

The availability of such information in practice and its truthfulness is always questionable as such information is available from the experience of the experimenter, the long association of the experimenter with the experiment, and similar types of studies conducted in the past for other external resources from outside the sample. The authors of Ref. [2, p. 163] have explicitly stated that "Frequently, the additional information needed for identification is either not available or is not widely shared by researchers in the field." In other interdisciplinary subjects also, such issues arose; for instance, see [3–5] and a few references therein. Keeping in mind the difficulties faced by the parametric procedures, another alternative is to use non-parametric procedures when the data is contaminated with measurement errors.

The non-parametric procedures are free from these types of drawbacks and limitations, and various estimation procedures have been explored in the literature for fitting the non-parametric models. In both non-parametric and parametric measurement error models, there have been a few works in the literature. For instance, [6] and [7] dealt with non-parametric regression in the presence of measurement error. Ref. [6] showed that, if the measurement error is normally distributed with a known variance having sample size n, no consistent non-parametric estimator of $m(.)$ converges faster than the rate $\{log\,(n)\}^{-k}$, where $m(.)$ is $k\,(\geq 1)$ times differentiable. Ref. [8] studied the impact of measurement errors on linear regression with the lasso penalty, and Ref. [9] presented the method of moments estimation for generalized linear measurement error models using the concept of instrumental variables. The authors of [10] have generalized the cumulative slicing estimator for dimension reduction, where the predictors are contaminated with measurement errors. Ref. [11] proposed a general estimating principle that is widely applicable when the score function can be estimated without bias.

To obtain a good model, various estimation procedures can be chosen essentially based on their statistical properties. A common user will be more interested in finding a better-fitted model to a given data than in the choice of estimation procedure, which can further be used for other purposes like forecasting, etc. Goodness of fit measures based on different information criteria, e.g., coefficient of determination, Akaike information criterion (AIC), Bayesian information criterion (BIC), etc., have been proposed in the literature mainly for situations when the data is free from measurement errors. For more information on theoretical approaches, readers may refer to [12–15] and the references therein. In this context, one may note that any goodness of fit statistic depends on the estimated parameters. It is indeed true that the properties of the estimators differ significantly based on whether the measurement errors are present or absent in the data. Consequently, any goodness of fit statistic developed for no measurement error situations is likely to become invalid for the situations when measurement error is present in the data. Furthermore, how to judge the goodness of fit of the model under such situations using a specified goodness of fit criteria is not much explored in the literature.

Recently, Ref. [16] proposed a measure to judge the goodness of fit of the fitted multiple parametric measurement error model but their proposed measures are dependent on additional information such as the known covariance matrix of measurement errors or known reliability matrix. However, how to measure the goodness of fit of the fitted multiple non-parametric model in the presence of measurement error has not yet been addressed in the literature, to the best of our knowledge. To inspect this problem, various well-known estimators like Nadaraya-Watson and Priestley-Chao estimators can be used to estimate the regression function, and one can propose to use them to develop the measure of goodness of fit for the multiple non-parametric measurement model. This gives rise to several questions which need to be answered and have not been addressed in the literature, to the best of our knowledge.

Aims and Issues:
We attempt to address the following issues in this paper.

  i. Among those possible choices of the measures of goodness of fit for the multiple non-parametric measurement error model, which estimator will perform well in terms of fitting a model to a given data.
 ii. Furthermore, one also would be interested to know, how to use these measures in variable selection problems for a given data.
iii. Finally, how good is the performance of the proposed measures of goodness of fit statistics or information criteria when the data will be generated from a complex model.

The modest objective of this article is to address issues described in (i), (ii), and (iii).

Traditionally, the coefficient of determination or R^2 information criteria is utilized to determine the degree of goodness of fit in classical multiple linear regression model when measurement errors are absent in the data. Such a R^2 statistic or information criteria is based on the partitioning of the total sum of squares into two orthogonal components, namely, the sum of squares due to regression and the sum of squares due to random errors. In other words, the R^2 information criteria essentially compute the proportion of variability in the data that is being explained by the fitted model in comparison to the total variation. Such partitioning of the sum of squares is not possible in case the data has measurement errors, and hence, the traditional R^2 cannot be used.

The idea and the philosophy of goodness of fit statistic from multiple linear regression models without measurement error motivated us to construct a goodness of fit statistic, which can measure the level of goodness of fit for the multiple non-parametric regression model with measurement error in covariates, which has not yet been available in the literature, to the best of our knowledge. As mentioned earlier in (i) and (ii), such a statistic can be used not only to measure the goodness of fit of a model to a given data but also for the variable selection process. In fact, as it is pointed out in (iii) in Section 1.1, we will see in Section 1.4 that such proposed information criteria is more useful than the usual R^2, when the data are generated from complex models; strictly speaking, which are much different from linear regression model. That is, the correct choice of appropriate regression function to specify the model on the basis of given data can also be recommended.

Further ideally, one would always be interested in finding the exact or asymptotic distribution of goodness of fit statistic using a general non-parametric estimator, but this is extremely difficult to derive. Even if derived, the expression may be so complicated that it may not shed light on the properties of the goodness of fit statistic. Moreover, the use of different estimators will give different forms of goodness of fit statistic. In order to illustrate the statistical implications, we propose to use Nadarya-Watson and Pristley-Chao estimators to estimate the unknown regression function in the presence of measurement errors in the covariates, and subsequently, they are used to obtain the goodness of fit statistics based on each of the estimator. The asymptotic distributions of the goodness of fit statistics based on these estimators are derived, and the practicability of the proposed goodness of fit statistics is shown in real data and simulation study.

The rest of the article is organized as follows. In Section 1.2, we describe the non-parametric measurement error model and propose the goodness of fit statistic under the model. The asymptotic distributions of the proposed two goodness of fit statistics are studied in Section 1.3. Section 1.4 investigates the performance of those goodness of fit statistics on different simulated data and real data as well. Some concluding remarks are discussed in Section 1.5. All technical details of the tests are provided in the Appendix.

1.2 MODEL DESCRIPTION AND PRELIMINARY CONCEPTS

1.2.1 Multiple non-parametric measurement error model

Consider the multiple non-parametric measurement error model in which the measurement errors are assumed to be present only in the covariates. The paired observations are available on the correctly observed response and $p-$dimensional measurement error-ridden covariates as $(y_1^*, \mathbf{x}_1^*)$, ..., $(y_n^*, \mathbf{x}_n^*)$ which are related with the model

$$y_i^* = m\left(\mathbf{x}_i^*\right) + \varepsilon_i, \; i = 1, ..., n, \tag{1.1}$$

$$\mathbf{x}_i^* = \mathbf{x}_i + \eta_i \tag{1.2}$$

where y_i^* is the i-th observed response variable, $\mathbf{x}_i^* = (x_{i1}^*, \ldots, x_{ip}^*)$ is the i-th observed p-dimensional ($p \geq 1$) covariate, and ε_i is the i-th variable of the random error in the model. In model (1.1), (1.2), $\eta_i = (\eta_{i1}, \ldots, \eta_{ip})$ is the p-dimensional measurement error involved in the i-th observed p-dimensional covariate $\mathbf{x}_i^*$, i.e., precisely speaking, if $\mathbf{x}_i = (x_{i1}, \ldots, x_{ip})$ is the i-th true covariate, and + sign indicates the componentwise addition of two vectors. The unknown function $m(.)$ in (1.1) is the non-parametric regression function which is a function of random η_i and true but unobservable $\mathbf{x}_i$ but we use the notation $m(.)$ due to notational simplicity. To summarize, this is the multiple non-parametric measurement error model, where the measurement error is involved in the covariate. A few more technical conditions will be assumed when the theoretical results are stated. In this case, not all the covariates are measured with measurement errors, then such a situation can be handled by substituting $\eta_i = 0$ for such covariates in the model, where $0 = (0, \ldots, 0)$ is a p-dimensional vector having each element equal to zero.

1.2.2 Goodness of fit in parametric regression models

Let us first explain the idea behind the goodness of fit statistic in a classical multiple linear regression model (without measurement error), which is measured by the coefficient of determination (R^2) information criteria. This explanation will help in understanding the development of goodness of fit measure for the multivariate non-parametric measurement error model. *The notations and content used in this subsection are limited to this subsection only.* A multiple linear regression model is described by

$$y = X\beta + u,$$

where $y = (y_1, \ldots, y_n)$ denotes a n-dimensional vector of observations on the response variable, X denotes the ($n \times (p + 1)$)-dimensional matrix of n observations on each of the p-dimensional covariate and an intercept term, $\beta = (\beta_0, \beta_1, \ldots, \beta_p)$ is the ($p + 1$)-dimensional vector of regression coefficients associated with the p-dimensional covariate and the intercept term, and $u = (u_1, \ldots, u_n)$ denotes the n-dimensional vector of random errors in the model with $E(u) = 0$ and $E(uu') = \sigma_u^2 I_n$. Suppose that $\hat{\beta}$ is the ordinary least squares estimator of the unknown parameter β, and it is well-known that $\hat{\beta} = (X'X)^{-1}X'y$ when $(X'X)^{-1}$ exists. The conventional information criteria, termed as coefficient of determination (denoted by R^2) in the multiple linear regression model is defined as the ratio of the explained variation to the total variation. The respective variations are measured in terms of sum of squares. Let *SSReg*, *SSError*, and *SSTotal* denote the sum of squares due to regression, errors and total, respectively in the multiple linear regression model. We then have

$$R^2_{conventional} = \frac{SSReg}{TotalSS} = \frac{SSReg}{SSError + SSReg}, \quad 0 \le R^2 \le 1,$$

where $SSReg = \hat{\beta}'X'X\hat{\beta}$, $SSError = (y - X\hat{\beta})'(y - X\hat{\beta})$, $TotalSS = \sum_{i=1}^{n}(y_i - \bar{y})^2$, and $\bar{y} = \frac{1}{n}\sum_{i=1}^{n} y_i$. Observe that the statistic or information criteria R^2 is the sample-based multiple correlation coefficient between y and the p-dimensional covariate. In fact, R^2 is a consistent estimator of the population multiple correlation coefficient (denoted as $\rho^2_{conventional}$) between the response variable and the multi-dimensional covariate, and it is used for measuring the goodness of fit of the model based on when the unknown parameters are estimated by the ordinary least squares estimation. The exact form of $\rho^2_{convensional}$ is the following.

$$\rho^2_{conventional} = \frac{\beta'\Sigma_{XX}\beta}{\beta'\Sigma_{XX}\beta + \sigma_u^2},$$

where $\Sigma_{XX} = \lim_{n\to\infty} n^{-1}X'PX$, $P = I_n - n^{-1}\mathbf{1}\mathbf{1}'$ and $\mathbf{1}$ is a n-vector of all the elements unity (see [17], Chapter 4).

It may be noted that in the context of analysis of variance (ANOVA) in the multiple linear regression model, *SSReg* and *SSError* are independently distributed. Thus, the *TotalSS* is partitioned into two mutually orthogonal components, namely, *SSReg* and *SSError*. Such an orthogonal partitioning is not possible if the basic assumptions of the multiple linear regression model are violated, and that is why the definition of goodness of fit (R^2) does not extend directly to such situations as non-parametric multivariate measurement error model directly. For example, when X and u are not independent in the multiple linear regression model, which happens if the covariate contains measurement error, the usual assumptions associated with ANOVA are violated. Consequently, the ordinary least squares estimator of β becomes a biased and inconsistent estimator of β, and hence, the conventional R^2 cannot be defined and used as the goodness of fit statistic in such a situation. Strictly speaking, the use of R^2 under the aforesaid situation will lead to incorrect statistical inferences. Similar situations occur in the case of multiple non-parametric measurement error models. This needs attention, and a possible solution can be achieved by looking at the fundamental problem that is disturbing the construction of R^2 information criteria when the model does not satisfy the conditions of usual ANOVA.

To illustrate this, we explain through the parametric set up of the measurement error model where the classical R^2 is not usable. The ordinary least squares estimator, i.e., $\hat{\beta}$, which is the best linear unbiased estimator of β in a multiple linear regression model with no measurement errors, becomes not only biased but also an inconsistent estimator of β under the measurement error models. The consistent estimators of regression coefficients in the multiple linear measurement error models are obtained by adjusting the ordinary least squares estimator for their inconsistency. Such an adjustment can be done by using the additional information from outside the sample. In the context of the multiple linear measurement error model, there are two possible forms of additional information, which can be used to obtain consistent estimators of the regression coefficient vector. These two forms are based on the knowledge of the covariance matrix of measurement errors associated with covariate and the knowledge of the reliability matrix of covariate, see, e.g., [18] and [19]. Cheng et al. [20] proposed two statistics for measuring the goodness of fit in the ultrastructural form of the multiple linear measurement error model, see also [16], where they provided the goodness of fit in a multiple linear measurement error model, when the regression coefficients are bounded by an exact linear restriction. Two versions of goodness of fit statistics proposed in articles [16] and [20] are based on the assumption of known covariance matrix of measurement errors and known matrix of reliability ratios. Such information may not be available in many practical situations, and this restricts the use of their proposed goodness of fit statistics. Moreover, the exact form of the relationship between X and β may be complicated and may not be accurately known in many situations. This motivates us to consider the multiple non-parametric measurement error model since such methods are free from the aforementioned limitations.

1.2.3 Goodness of fit in non-parametric measurement error model

Motivated by the formulation of R^2 information criteria in the multiple linear regression model, we try here to develop analogous goodness of fit statistics for multiple non-parametric measurement error models. Recall the model described in (1.1), and let the unknown regression function $m(.)$ be estimated by $\hat{m}_n(.)$ based on the measurement error-ridden observed data $(y_1^*, \mathbf{x}_1^*), ..., (y_n^*, \mathbf{x}_n^*)$. Here $m(.): A \to \mathbb{R}$, where A is any arbitrary bounded subset of $\mathbb{R}^p (p \geq 1)$. Now, for any given $\mathbf{z} = (z_1, ..., z_p) \in A$, define

$$R_n^{*2}(z) = \frac{\hat{m}_n^2(\mathbf{z})}{\sigma_\varepsilon^2 + \hat{m}_n^2(\mathbf{z})},$$

where σ_ε^2 is the variance of the random errors ε (due to model). However, in practice, σ_ε^2 is unknown, and hence, for any $\mathbf{z} \in A$, the expression $R_n^{*2}(\mathbf{z})$ is not computable for a given data. In order to make it computable, one can estimate σ_ε^2 by any of its consistent estimators, and one such possible estimator is

$$\widehat{\sigma_{\varepsilon,n}^2} = \frac{1}{n-1} \sum_{i=1}^{n} (y_i^* - \hat{m}_n(x_i^*))^2.$$

Using $\widehat{\sigma_{\varepsilon,n}^2}$, for any $\mathbf{z} \in A$, let us denote

$$R_n^2(\mathbf{z}) = \frac{\hat{m}_n^2(\mathbf{z})}{\widehat{\sigma_{\varepsilon,n}^2} + \hat{m}_n^2(\mathbf{z})}, \tag{1.3}$$

which can apparently be thought of as a possible choice of the goodness of fit statistic in the multiple non-parametric measurement error model for the following reasons:

Observe that for any $\mathbf{z} \in A$, $R_n^2(\mathbf{z}) \in [0,1]$, which gives a preliminary impression that $R_n^2(\mathbf{z})$ can be a possible choice of goodness of fit statistic. Moreover, note that $R_n^2(\mathbf{z}) \overset{a.s}{\to} 0$ as $n \to \infty$ when $\hat{m}_n^2(\mathbf{z}) \overset{a.s}{\to} 0$ as $n \to \infty$ for any given $\mathbf{z} \in A$, where $a.s$ stands for almost sure convergence. This fact implies that for a given $\mathbf{z} \in A$, the measure $R_n^2(.)$ will converge to zero when the fitted models fails to have relevant covariates responsible for explaining the variation in the observations through the functional relationship between the response variable and the covariate for a sufficiently large sample size, which is expected to be true. At the same time, one can see that $R_n^2(\mathbf{z}) \overset{a.s}{\to} 1$ as $n \to \infty$ for a given $\mathbf{z} \in A$ when $\widehat{\sigma_{\varepsilon,n}^2} \overset{a.s}{\to} 0$ as $n \to \infty$. This fact indicates that for a given $\mathbf{z} \in A$, the measure $R_n^2(.)$ attains its maximum value one when there is no variability among the errors due to regression for a sufficiently large sample, and in turn, the fitted model is perfect. This is also seemingly accordant with the idea of the goodness of fit statistic. However, note that the aforesaid statements depend on the value of the given $\mathbf{z}$. For instance, for a given data, it may happen that $R_n^2(\mathbf{z}_1) = 0.71$ when $\mathbf{z} = \mathbf{z}_1$ whereas for $\mathbf{z} = \mathbf{z}_2$, $R_n^2(\mathbf{z}_2) = 0.97$, i.e., strictly speaking, for the same data, the various choices of $\mathbf{z}$ lead to different values of $R_n^2(\mathbf{z})$. This issue makes the use of $R_n^2(\mathbf{z})$ improper and inappropriate for measuring the goodness of fit in the multiple non-parametric measurement error model. To overcome this problem, we modify $R_n^{*2}(\mathbf{z})$ and propose the goodness of fit statistics in the following way.

It is clear that one needs to accumulate the information of $R_n^2(\mathbf{z})$ for all possible choices of $\mathbf{z} \in A$ to solve this issue. In order to have all information, we modify and propose the goodness of fit statistic, i.e., R^2 information criteria in the multiple non-parametric measurement model as

$$R_n^2 = \frac{\int_A R_n^2(\mathbf{z}) \, d\mathbf{z}}{\Lambda(A)}, \tag{1.4}$$

where $\Lambda(A)$ denotes the Lebesgue measure of A in $\mathbb{R}^p$ ($p \geq 1$). First note that $R_n^2 \in [0,1]$ for all $n \in \mathbb{Z}$. Next, it follows from the definition of R_n^2 that this information criteria does not depend on any particular choice of $\mathbf{z} \in A$ since it is integrating "all" information of $R_n^2(\mathbf{z})$ over "all" $\mathbf{z} \in A$, and hence, for a given data, R_n^2 will provide us a single value to determine the goodness of fit of the multiple non-parametric measurement error model. It is appropriate to mention that the population version of R_n^2 is

$$\rho^2 = \frac{\int_A \frac{m^2(z)}{\sigma_\varepsilon^2 + m^2(z)} \, dz}{\Lambda(A)}, \tag{1.5}$$

Here also, note that $\rho^2 \in [0, 1]$. Further, note that the values of R_n^2 depend upon the integration, which can be solved using numerical integration techniques with the help of software such as R, Matlab, Mathematica, etc., or by various methodologies of Monte Carlo Simulation.

Now we discuss how R_n^2 can address the issues mentioned in (i) and (ii) in the Introduction. To address (i), let $R_{1,n}^2$ and $R_{2,n}^2$ be two versions of R_n^2 to check the validity of a particular model to fit a given data. In this case, one may prefer the version of which one will give larger value and carry out the entire analysis based on the respective version of R_n^2. This concept can be true for more than two versions of R_n^2 also. To address (ii), i.e., related to variable selection problem, suppose that M_1, M_2, and M_3 are three candidate models with m_1, m_2, and m_3 number of variables, respectively. Among these three choices, one can prefer m_2 number of variables over m_1 and m_3 number of variables when the value R_n^2 for m_2 number of variables is more than the values of R_n^2 for m_1 and m_3 number of variables. This concept can also be extended for any finite number of variables. The usefulness of R_n^2 in addressing (i) and (ii) in Section 1.1 will be demonstrated for simulated and real data also in Sections 1.4.1 and 1.4.2.

1.3 LARGE SAMPLE STATISTICAL PROPERTIES OF R_n^2

It is clear from the definition of R_n^2 in (1.4) and the expression of $R_n^2(\mathbf{z})$ in (1.3) that the distributional feature of R_n^2 depends on the distribution of $m(.)$, which is an estimator of the unknown non-parametric regression function $m(.)$, and hence, various choices of $\hat{m}_n$ leads to different distributions of the R^2 information criteria in the multiple non-parametric measurement error model. In this article, well-known Nadaraya-Watson estimator (see, [21] and [22]) and Priestley-Chao estimator (see, [23]) are considered as $\hat{m}_n$, which are denoted as $\hat{m}_n^{NW}(.)$ and $\hat{m}_n^{PC}(.)$, respectively. The forms of $\hat{m}_n^{NW}(.)$ and $\hat{m}_n^{PC}(.)$ are as follows. For any $\mathbf{z} \in A$,

$$\hat{m}_n^{NW}(\mathbf{z}) = \sum_{i=1}^{n} y_i^* K\left(\frac{\mathbf{z} - \mathbf{x}_i^*}{b_n}\right) \bigg/ \sum_{i=1}^{n} K\left(\frac{\mathbf{z} - \mathbf{x}_i^*}{b_n}\right) \tag{1.6}$$

and

$$\hat{m}_n^{PC}(\mathbf{z}) = \sum_{i=1}^{n} y_i^* K\left(\frac{\mathbf{z} - \mathbf{x}_i^*}{b_n}\right) \bigg/ nb_n^p, \tag{1.7}$$

where $K(.)$ is the multivariate kernel function, and $\{b_n\}$ is a sequence of bandwidth parameters. It is an appropriate place to mention that the kernel $K(.)$ is useful to construct the estimator of a probability density function, and $\{b_n\}$ is a sequence of bandwidth associated with the kernel $K(.)$. For details on the kernel density estimation and the associated bandwidth sequence $\{b_n\}$, readers may refer to Chapters 3 and 4 in [24]. Let $R_{n,NW}^2$ and $R_{n,PC}^2$ denote the information criteria R_n^2 based on $\hat{m}_n^{NW}(.)$ and $\hat{m}_n^{PC}(.)$, respectively, i.e.,

$$R_{n,NW}^2 = \frac{\int_A \frac{\{\hat{m}_n^{NW}(\mathbf{z})\}^2}{\sigma_{\varepsilon,n}^2 + \{\hat{m}_n^{NW}(\mathbf{z})\}^2} d\mathbf{z}}{\Lambda(A)} \tag{1.8}$$

and

$$R_{n,PC}^2 = \frac{\int_A \frac{\{\hat{m}_n^{PC}(\mathbf{z})\}^2}{\sigma_{\varepsilon,n}^2 + \{\hat{m}_n^{PC}(\mathbf{z})\}^2} d\mathbf{z}}{\Lambda(A)}. \tag{1.9}$$

As mentioned earlier, $\Lambda(A)$ denotes the Lebesgue measure of set A in $\mathbb{R}^p$ ($p \geq 1$). This section studies the asymptotic properties of $R_{n,NW}^2$ and $R_{n,PC}^2$, and before stating the main results, one needs to assume the following conditions on the multiple non-parametric measurement error model described in Section 1.2.1.

A1. The true values of the covariates $\mathbf{x}_i$ ($i = 1, \ldots, n$) are bounded. In other words, there exists some M_1 such that for $i = 1, \ldots, n$, $\|\mathbf{x}_i\| \leq M_1$, where $\|.\|$ denotes the Euclidean norm.

A2. The measurement errors $\eta_i = (\eta_{i1}, ..., \eta_{ip})$ $(i = 1, ..., n)$ are i.i.d. sequence of p-dimensional $(p \geq 1)$ bounded and continuous random vectors, i.e., there exists some M_2 such that $\|\eta_i\| \leq M_2$ $(i = 1, ..., n)$. Moreover, the probability density function of η_i $(i = 1, ..., n)$ is a bounded and twice differentiable function. Furthermore, $E(\eta_i) = 0 = (0, ..., 0) \in \mathbb{R}^p$ $(i = 1, ..., n)$ and $E(\eta_{ik}\eta_{il}) = 0$ for all $1 \leq k \neq l = p$ and $i = 1, ..., n$.

A3. The regression errors ε_i $(i = 1, ..., n)$ are i.i.d. sequence of random variables with $E(\varepsilon_i) = 0$ and $E(\varepsilon_i^2) = \sigma_\varepsilon^2 < \infty$. Moreover, for $i = 1, ..., n$, $E(\varepsilon_i\eta_{ik}) = 0$ for all $k = 1, ..., p$.

A4. For $p \geq 1$, $K \colon \mathbb{R}^p \to \mathbb{R}$ is centrally symmetric about $0 = (0, ..., 0) \in \mathbb{R}^p$, i.e., $K(-\mathbf{a}) = K(\mathbf{a})$ for all $\mathbf{a} \in \mathbb{R}^p$. Moreover, K is satisfying $\int uK(u)\,du = 0$ and $\int u'uK(u)\,du < \infty$.

A5. *The sequence of bandwidth $\{h_n\}$ is such that $h_n \to 0$ as $n \to \infty$, and $nh_n^p \to \infty$ as $n \to \infty$.*

A6. *$m(.)$ is thrice continuously differentiable function on the bounded set $A \subset \mathbb{R}^p$, where A is the same as defined in (1.4).*

Remark 1: Note that (A1) is a realistic condition in a regression problem. Condition (A2) indicates that the measurement errors cannot be unbounded, which is clearly true in real data applications, else no statistical tool will remain valid. Besides, twice differentiability of the probability density functions of the measurement error random variables satisfies for many probability density functions. Moreover, condition (A2) implies that the components of measurement error random vectors are uncorrelated, which is also expected in practice. Condition (A3) is a common assumption in the literature of regression analysis and measurement error models, and it is realistic as well. Condition (A4) holds for most of the well-known kernel density functions (see, e.g., [24]), and condition (A5) satisfies many choices of the bandwidth h_n. Finally, many choices of m satisfy condition (A6). This ensures that the proposed goodness of fit statistic/information criteria can be used in real data applications without any complications arising due to theoretical constraints.

Remark 2: Observe that conditions (A1) and (A2) imply that set A in the expression of R_n^2 (see (1.4)) is a bounded subset of $\mathbb{R}^p$ $(p \geq 1)$. Precisely speaking, $A = \{\mathbf{a} \in \mathbb{R}^p \colon \|\mathbf{a}\| \leq M_1 + M_2\}$. Note that the volume of set A, i.e., $\Lambda(A)$ is finite since M_1 and M_2 are finite, defined in conditions (A1) and (A2), respectively. This fact implies that the definitions of R_n^2 in (1.4) and ρ^2 in (1.5) are well-defined.

The following theorems state the asymptotic distributions of $R_{n,NW}^2$ defined in (1.8) and $R_{n,PC}^2$ defined in (1.9).

Theorem 1.1: *Under conditions (A1)–(A6),* $\sqrt{nb_n^p}\left(R_{n,NW}^2 - \rho^2 - \int_A \frac{P(\mathbf{z})}{\Lambda(A)}d\mathbf{z}\right)$ *converges weakly to the distribution of* $\int_A \dfrac{\sigma_\varepsilon^2 T_1(\mathbf{z})}{\Lambda(A)(\sigma_\varepsilon^2 + \{m(\mathbf{z}) + \{B(\mathbf{z})\}^2)\left(\sigma_\varepsilon^2 + \{m(\mathbf{z})\}^2\right)}d\mathbf{z},$

where A is the same as defined in Remark 2, $\Lambda(A)$ *is the Lebesgue measure of A in* $\mathbb{R}^p$, $E(\varepsilon_i^2) = \sigma_\varepsilon^2$ *for* $i = 1, \ldots, n$, ρ^2 *is the same as defined in (1.5), and* $T_1(z)$ *is a random variable associated with normal distribution with mean* $= a_1(\mathbf{z})$ *and variance* $= b_1(\mathbf{z})$. *The expressions of* $a_1(\mathbf{z})$, $b_1(\mathbf{z})$, $P(\mathbf{z})$, *and* $B(\mathbf{z})$ *are provided in the proof of the theorem.*

Theorem 1.2: *Under conditions (A1)–(A6),* $\sqrt{nb_n^p}\left(R_{n,PC}^2 - \rho^2 - \int_A \frac{Q(\mathbf{z})}{\Lambda(A)}d\mathbf{z}\right)$ *converges weakly to the distribution of*

$$\int_A \frac{\sigma_\varepsilon^2 T_2(\mathbf{z})}{\Lambda(A)\left(\sigma_\varepsilon^2 + \left(m(\mathbf{z})\int_A f_\eta(\mathbf{z}-y)dy + \tilde{B}(\mathbf{z})\right)^2\right)(\sigma_\varepsilon^2 + \{m(\mathbf{z})\}^2)}d\mathbf{z}, \text{ where A is the same as defined}$$

in Remark 2, $\Lambda(A)$ *is the Lebesgue measure of A in* $\mathbb{R}^p$, $E(\varepsilon_i^2) = \sigma_\varepsilon^2$ *for* $i = 1, \ldots, n$, ρ^2 *is the same as defined in (1.5),* f_η *is the probability density function of* η, *and* $T_2(\mathbf{z})$ *is a random variable associated with normal distribution with mean* $= a_2(\mathbf{z})$ *and variance* $= b_2(\mathbf{z})$. *The expressions of* $a_2(\mathbf{z})$, $b_2(\mathbf{z})$, $Q(\mathbf{z})$ *and* $\tilde{B}(\mathbf{z})$ *are provided in the proof of the theorem.*

Remark 3: The assertions in Theorems 1 and 2 imply that $R_{n,NW}^2 - \int_A \frac{P(\mathbf{z})}{\Lambda(A)}d\mathbf{z}$ and $R_{n,PC}^2 - \int_A \frac{Q(\mathbf{z})}{\Lambda(A)}d\mathbf{z}$ converge in probability to ρ^2 (see (1.5)), which is the population version of R_n^2 information criteria in the multiple non-parametric measurement error model described in Section 1.2.1. As $\int_A P(\mathbf{z})d\mathbf{z}$ and $\int_A Q(\mathbf{z})d\mathbf{z}$ are unknown in practice, one needs to estimate these two expressions from the given data to implement the results of Theorems 1 and 2 on any real data sets.

1.4 FINITE SAMPLE STUDY AND REAL DATA ANALYSIS

1.4.1 Finite sample study

In Section 1.3, we studied the large sample properties of $R_{n,NW}^2$ and $R_{n,PC}^2$. We now want to see the performance of the proposed information criteria, i.e., $R_{n,NW}^2$ and $R_{n,PC}^2$ for finite sample simulated data and real data. We will here address the issues (i), (ii), and (iii) in Section 1.1 along with the study on the performance of the proposed information criteria when the covariates do not have the measurement errors. This will give us an idea how the proposed goodness of fit criteria works for fitting a multiple non-parametric regression model with the measurement errors in the covariates.

1.4.1.1 Addressing (iii) mentioned in Section 1.1

Recall the regression model (1.2.1) and consider the following examples.

Example 1: $m(\mathbf{z}) = \sum_{i=1}^{p} a_i z_i$

Example 2: $m(\mathbf{z}) = \sum_{i=1}^{p} a_i z_i^2$

Example 3: $m(\mathbf{z}) = \sum_{i=1}^{p} a_i \sin z_i$.

Here $\mathbf{z} = (z_1, \ldots, z_p)$ is a generic notation, and $a_i(1, \ldots, p)$ is the ith unknown parameter in the regression model. We considered various choices of the parameters and sample sizes for conducting the Monte Carlo experiment, and the conclusions from all the settings were similar. Therefore, for sake of space, we are reporting the results from only a selected setting. We conducted here a Monte Carlo simulation study with the following choices of parameters: $p = 3$, $n = 50$, $a_1 = 0.3$, $a_2 = 0.4$, and $a_3 = 0.5$. Note that a_1, a_2, and a_3 can choose any values in three different examples but to avoid notational complexity and to understand the behaviour of the proposed goodness of fit and its dependence on the choice of the form of $m(\mathbf{z})$, we consider the same (a_1, a_2, a_3) in Examples 1, 2, and 3. In the numerical study, the true covariates $\mathbf{x}_i = (x_{i1}, x_{i2}, x_{i3})$ $(i = 1, \ldots, 50)$ are the componentwise equidistant values on $[0,2] \times [0,2] \times [0,2]$. The realized values of measurement error random variable $\eta_i = (\eta_{i1}, \eta_{i2}, \eta_{i3})$ $(i = 1, \ldots, 50)$ are generated from a uniform distribution on $[0,2] \times [0,2] \times [0,2]$, and the errors due to regression, i.e., ε_i $(i = 1, \ldots, 50)$ are generated from a standard normal distribution. Observe that here $A = [0,4] \times [0,4] \times [0,4]$ (see Remark 2 for details on the exact form of A), and consequently, $\Lambda(A) = 4^3 = 64$, where $\Lambda(.)$ denotes the Lebesgue measure. Throughout the simulation study, we consider three fold products of Epanechnikov kernel (see, e.g., [24]) as the kernel function unless mentioned otherwise, and $h_n = constant. \, n^{-1/5}$, where $constant$ is estimated by well-known cross-validation technique (see, e.g., [25]). Using the aforesaid information and the form of the model described in (1.1), for each example (i.e., Examples 1, 2, and 3), we generated the three sets of observed data with size 50, which is denoted as $(y_1^*, \mathbf{x}_1^*), \ldots, (y_{50}^*, \mathbf{x}_{50}^*), \mathbf{x}_{50}^*)$, where $\mathbf{x}_i^* = \mathbf{x}_i + \eta_i$, $i = 1, \ldots, 50$. We replicate this generation procedure of the observed data for 5000 times.

For each example, let $R_{conventional,k}^2$, $R_{n,NW,k}^2$ and $R_{n,PC,k}^2$ be the values of $R_{conventional}^2$, $R_{n,NW}^2$ and $R_{n,PC}^2$, respectively for k-th replicated observed data, where $k = 1, \ldots, 5000$. Next, for each example, we compute the empirical mean squares error (EMSE) of $R_{conventional}^2$, $R_{n,NW}^2$ and $R_{n,PC}^2$, which are defined as $\frac{1}{5000} \sum_{k=1}^{5000} (R_{conventional,k}^2 - \rho_{conventional}^2)^2$, $\frac{1}{5000} \sum_{k=1}^{5000} (R_{n,NW,k}^2 - \rho^2)^2$ and $\frac{1}{5000} \sum_{k=1}^{5000} (R_{n,PC,k}^2 - \rho^2)^2$, respectively. Here $\rho_{conventional}^2$ is the same as defined in Section 1.2.2, and ρ^2 is the same as defined in (1.5), and the results are the following. Intentionally, the data with higher values of $\rho_{conventional}^2$ is considered to avoid the obvious model defects.

For Example 1, we obtain $\rho^2_{conventional} = 0.659$ and $\rho^2 = 0.852$. Using these values of $\rho^2_{conventional}$ and ρ^2, we have EMSE $(R^2_{conventional}) = 0.114$ whereas EMSE $\left(R^2_{n,NW}\right) = 0.061$ and EMSE $\left(R^2_{n,PC}\right) = 0.057$. Observe that the value of EMSE $(R^2_{conventional})$ is larger than the values of EMSE $\left(R^2_{n,NW}\right)$ and EMSE $\left(R^2_{n,PC}\right)$, but the order of differences are not too large as EMSE $(R^2_{conventional})$/EMSE $\left(R^2_{n,NW}\right) = 1.86$ and EMSE $(R^2_{conventional})$/EMSE $\left(R^2_{n,PC}\right) = 2$. The reason is as follows: On the one hand, since the observed data are generated from a linear regression model, $R^2_{conventional}$ as an information criteria has advantage over $R^2_{n,NW}$ and $R^2_{n,PC}$. On the other hand, since the observed covariates have measurement error, as an information criteria, $R^2_{n,NW}$ and $R^2_{n,PC}$ have advantage over $R^2_{conventional}$. These two aforementioned facts lead to the obtained result of $R^2_{conventional}$, $R^2_{n,NW}$ and $R^2_{n,PC}$ for Example 1. Strictly speaking, even for linear regression with the covariates having measurement errors, $R^2_{n,NW}$ and $R^2_{n,PC}$ perform marginally better than $R^2_{conventional}$.

For Example 2, we obtain $\rho^2_{conventional} = 0.426$ and $\rho^2 = 0.910$. Using these values of $\rho^2_{conventional}$ and ρ^2, we have EMSE $(R^2_{conventional}) = 0.254$ whereas EMSE $\left(R^2_{n,NW}\right) = 0.007$ and EMSE $\left(R^2_{n,PC}\right) = 0.002$. Observe that the values of EMSE $\left(R^2_{n,NW}\right)$ and EMSE $\left(R^2_{n,PC}\right)$ are much lower than EMSE $(R^2_{conventional})$ as EMSE $(R^2_{conventional})$/EMSE $\left(R^2_{n,NW}\right) = 36.29$ and EMSE $(R^2_{conventional})$/EMSE $\left(R^2_{n,PC}\right) = 127$, i.e., in other words, the proposed information criteria $R^2_{n,NW}$ and $R^2_{n,PC}$ perform better than $R^2_{conventional}$. The reason is as follows: Since the observed data are generated from a non-linear regression model (strictly speaking, quadratic regression model), and the observed covariates have measurement errors, $R^2_{n,NW}$ and $R^2_{n,PC}$ are performing better than $R^2_{conventional}$. In this context, recall that the $R^2_{conventional}$, as an information criteria or goodness of fit statistic, only measures the linear relationship between the response variable and the covariates, and it performs well when there is no measurement error whereas the proposed information criteria $R^2_{n,NW}$ and $R^2_{n,PC}$ are developed for measuring the goodness of fit of any kind of functional relationship between the response variable and the covariates.

For Example 3, we obtain $\rho^2_{conventional} = 0.184$ and $\rho^2 = 0.620$. Using these values of $\rho^2_{conventional}$ and ρ^2, we have EMSE $(R^2_{conventional}) = 0.177$ whereas EMSE $\left(R^2_{n,NW}\right) = 0.0012$ and EMSE $\left(R^2_{n,PC}\right) = 0.00079$. Observe that the values of EMSE $\left(R^2_{n,NW}\right)$ and EMSE $\left(R^2_{n,PC}\right)$ are lower than EMSE $(R^2_{conventional})$ to a great extent as EMSE $(R^2_{conventional})$/EMSE $\left(R^2_{n,NW}\right) = 147.66$ and EMSE $(R^2_{conventional})$/EMSE $\left(R^2_{n,PC}\right) = 224.30$, in other words, for this example also, the proposed information criteria $R^2_{n,NW}$ and $R^2_{n,PC}$ perform much better than $R^2_{conventional}$ for Example 2. The reason is as follows: Since the observed data are generated from a sinusoidal regression model, which is different from the linear regression model to a large extent, and the observed covariates have measurement errors, $R^2_{n,NW}$ and $R^2_{n,PC}$ are performing better than $R^2_{conventional}$ to a large amount. Overall, it is clear from this study on Examples 1, 2, and 3, our proposed information criteria $R^2_{n,NW}$ and $R^2_{n,PC}$ outperform conventional $R^2_{conventional}$ when the data are associated with a complicated model, which is much dissimilar to the linear regression model, and the covariates have measurement errors.

1.4.1.2 Addressing (ii) mentioned in Section 1.1

In Section 1.4.1.1, we have seen that for Examples 2 and 3, the performances of $R^2_{n,NW}$ and $R^2_{n,PC}$ are better than the performance of $R^2_{conventional}$ as the models in Examples 2 and 3 are non-linear, and the covariates have measurement error in the models described in Examples 1, 2, and 3. Here we want to show how $R^2_{n,NW}$ can be useful in variable selection problems also. Note that one can use $R^2_{n,PC}$ also but for the sake of concise presentation, we choose only $R^2_{n,NW}$ in addressing the issue described in (ii) in Section 1.1. Suppose that in Example 2, p is either equal to three (i.e., $p = 3$) or equal to four (i.e., $p = 4$). In the course of generating observed data, for $p = 3$, we consider $a_1 = 0.3$, $a_2 = 0.4$, and $a_3 = 0.5$ as we chose in Section 1.4.1.1, and for $p = 4$, we consider $a_1 = 0.3$, $a_2 = 0.4$, $a_3 = 0.5$, and $a_4 = 0.6$. Next, following the same procedure as described in Section 1.4.1.1, we generate a data with size 50 from the model described in Example 2 for $p = 3$ and $p = 4$. Afterwards, for both data sets, we compute $R^2_{n,NW}$ and obtain $R^2_{n,NW} = 0.8895$ for $p = 3$, and $R^2_{n,NW} = 0.9528$ when $p = 4$. These values indicate that $R^2_{n,NW}$ is more when $p = 4$, and this concludes that $p = 4$ should be the appropriate number of variables for the given data. Indeed, this study can be done for more than two choices of p as well.

1.4.1.3 Addressing (i) mentioned in Section 1.1

As we have seen in Section 1.4.1.2, for the given data generated from the model in Example 2, $p = 4$ is selected based on the information criteria $R^2_{n,NW}$. As mentioned in Section 1.4.1.2, one can use $R^2_{n,PC}$ as well to solve the variable selection problem described in that section. Suppose that we would now like to know which one between $R^2_{n,NW}$ and $R^2_{n,PC}$ is more useful for this data (i.e., the observed data for $p = 3$) as a goodness of fit statistic or information criteria to measure how good is the non-parametric functional relationship between the response variable and the covariates.

To solve this problem, we compute both $R^2_{n,NW}$ and $R^2_{n,PC}$ for this data, and we obtain $R^2_{n,NW} = 0.9244$ and $R^2_{n,PC} = 0.9668$, i.e., $R^2_{n,NW} \leq R^2_{n,PC}$. Hence, for this observed data, our recommendation is to use $R^2_{n,PC}$ information criteria to measure the goodness of non-parametric regression model to fit the data.

1.4.1.4 Covariates not having measurement errors

Here we would like to study the performance of $R^2_{n,NW}$, $R^2_{n,PC}$, and $R^2_{conventional}$ when the data do not have any measurement error. To address this issue, we compute EMSE $(R^2_{n,NW})$, EMSE $(R^2_{n,PC})$, and EMSE $(R^2_{conventional})$ for Examples 2 and 3 for $p = 3$, when $\eta_i = (\eta_{i1}, \eta_{i2}, \eta_{i3}) = (0,0,0)$. For Example 2, we obtain EMSE $(R^2_{n,NW}) = 0.0056$, EMSE$(R^2_{n,PC}) = 0.0001$, and EMSE $(R^2_{conventional}) = 0.0513$, and for Example 3, EMSE $(R^2_{n,NW}) = 0.0015$, EMSE$(R^2_{n,PC}) = 0.0004$ and EMSE$(R^2_{conventional}) = 0.0246$. It is clear from these EMSE values that the proposed information criteria R^2_{NW} and $R^2_{n,PC}$ here also perform better than $R^2_{conventional}$ since the model in Examples 2 and 3 are much different from the linear model.

1.4.2 Real data analysis

1.4.2.1 Pig data

This data set was collected by the Statistical Laboratory of Iowa State University under contract to the Statistical Reporting Service, Department of Agriculture, United States of America. This data was earlier analysed in Battese, Fuller, and Hickman [26], and in the context of linear regression measurement error model, it was re-analysed in Fuller [27]. This data set has two variables, namely, the number of breeding hogs denoted as X^*, i.e., the observed covariate and the number of sows that have completed the farrowing process denoted by Y^*, i.e., the observed response variable, and the size (denoted as n) of the data set is 184. Observe that here $p = 1$, and the data is plotted in Figure 1.1. Since this data is collected by interviewing farmers, it is believed that the data may have measurement errors as mentioned in [27].

From this data set, we generate $B = 5000$ many Bootstrap re-samples with the same size $n = 184$, and compute the Bootstrap mean squares error (denoted as BMSE) of $R^2_{conventional}$, $R^2_{n,NW}$ and $R^2_{n,PC}$, which are defined as follows. Let $R^2_{conventional,obs}$, $R^2_{n,NW,obs}$ and $R^2_{n,PC,obs}$ be the values $R^2_{conventional}$, $R^2_{n,NW}$ and $R^2_{n,PC}$ for this data, respectively, and $R^2_{conventional,j}$, $R^2_{n,NW,j}$ and $R^2_{n,PC,j}$ be the values of $R^2_{conventional}$, $R^2_{n,NW}$ and $R^2_{n,PC}$ for j-th re-sample, where $j = 1, ..., B$. The BMSE of $R^2_{conventional}$, $R^2_{n,NW}$ and $R^2_{n,PC}$ are defined as

The Pig Data

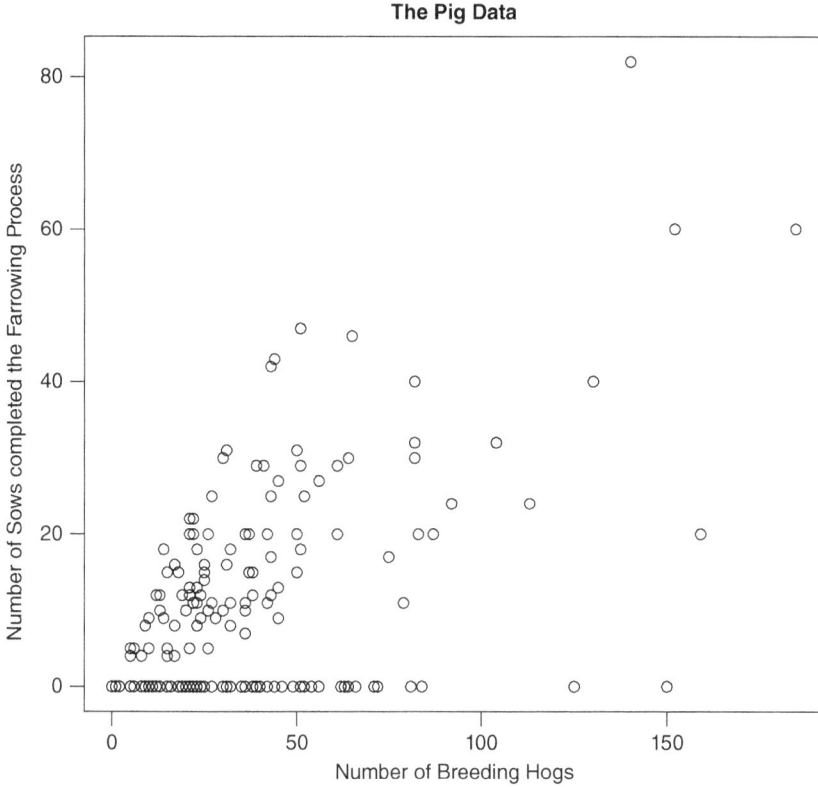

Figure 1.1 Scatter plot of pig data.

$\frac{1}{B} \Sigma_{j=1}^{B} (R^2_{conventional,j} - R^2_{conventional,obs})^2$, $\frac{1}{B} \Sigma_{j=1}^{B} (R^2_{n,NW,j} - R^2_{n,NW,obs})^2$ and $\frac{1}{B} \Sigma_{j=1}^{B} (R^2_{n,PC,j} - R^2_{n,PC,obs})^2$, respectively. For this data, we obtain BMSE $(R^2_{conventional}) = 0.0788$, BMSE$(R^2_{n,NW}) = 0.0009$ and BMSE$(R^2_{n,PC}) = 0.0003$. These BMSE values clearly indicate that the proposed information criteria $R^2_{n,NW}$ and $R^2_{n,PC}$ perform better than $R^2_{conventional}$, and the most probable reason is that here the response variable and the covariate has a complicated functional relationship, which is indicated from Figure 1.1.

1.4.2.2 Tourism data

The data set was collected from the Romanian Statistical Year Book 2008, which is available at https://insse.ro/cms/en/content/statistical-yearbooks-romania. In the context of the measurement error model, this data set was earlier analysed in Kulcsar [28] under linear regression model. This data has two variables, namely, investments in hotels and restaurants (denoted as X^*), i.e., the observed covariate, and the GDP obtained in the sector of hotels and

The Tourism Data

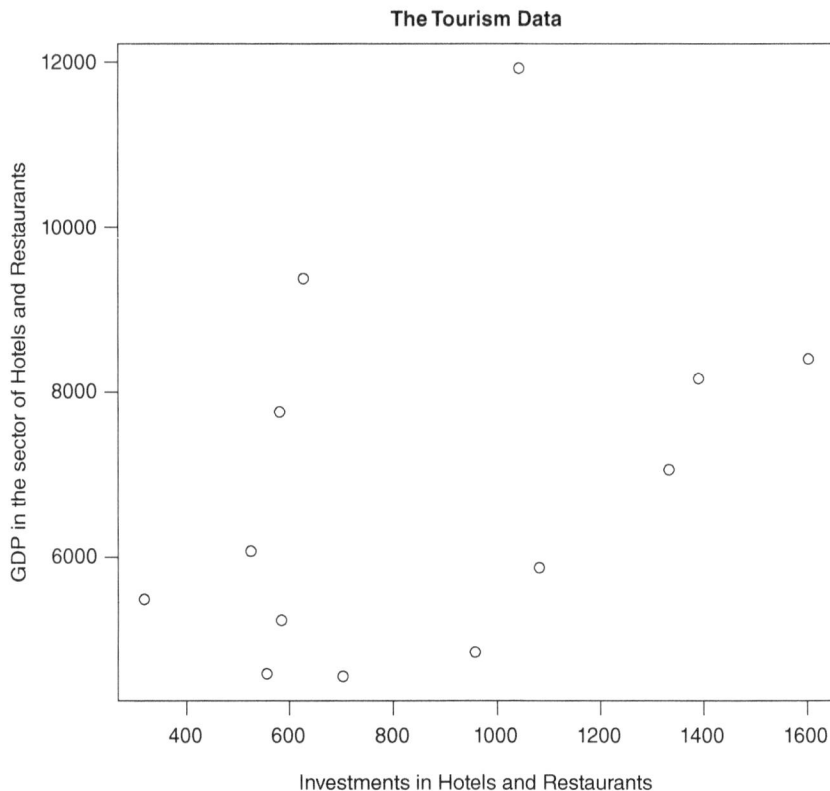

Figure 1.2 Scatter plot of tourism data.

restaurants (denoted as Y^*), i.e., the observed response variable, and the size of the data set is 13. Here also, observe that $p = 1$, and the data is plotted in Figure 1.2.

For this data also, we compute $\mathrm{BMSE}(R^2_{conventional})$, $\mathrm{BMSE}(R^2_{n,NW})$ and $\mathrm{BMSE}(R^2_{n,PC})$ as we did in Section 1.4.2.1, and we obtain $\mathrm{BMSE}(R^2_{conventional}) = 0.0517$, $\mathrm{BMSE}(R^2_{n,NW}) = 0.0026$ and $\mathrm{BMSE}\ (R^2_{n,PC}) = 0.0002$. These BMSE values indicate that the proposed information criteria $R^2_{n,NW}$ and $R^2_{n,PC}$ perform better than the conventional information criteria $R^2_{conventional}$. The superior performance of the proposed information criteria is explainable since it is clear from the plot in Figure 1.2 that Y^* and X^* has a non-linear relationship. Moreover, the presence of measurement error in the covariate also leads to an inferior performance of the conventional information criteria $R^2_{conventional}$. Overall, the conclusion is that once the data are generated from a model, far different than linear model, our proposed information criteria $R^2_{n,NW}$ and $R^2_{n,PC}$ are useful in the presence of measurement error in the covariates.

1.5 CONCLUDING REMARKS

In this article, we propose new information criteria or goodness of fit statistic to measure how good a multiple non-parametric measurement error model is in fitting a data set. Our proposed information criteria is based on the well-known Nadaraya-Watson and Priestley-Chao estimators of non-parametric regression function. We have seen that the proposed information criteria perform well when the data are generated from a complicated stochastic model, where "complicated stochastic model" refers to those models which are much different from linear models. Moreover, in the presence of measurement error also, the proposed information criteria have performed well. To summarize, strictly speaking, the proposed information criteria perform better than the conventional information criteria for any non-parametric regression model with or without measurement errors and for even linear regression models when the covariates have measurement errors.

The choice of kernel is also an issue of interest. As mentioned in Section 1.4, the threefold products of well-known Epanechnikov kernel are used in the numerical study since Epanechnikov kernel density estimator is the most efficient estimator among all symmetric kernel density estimators (see, e.g., [24], p. 43). However, in principle, one may use other kernel density estimators as well in practice, and the performance of the proposed information criteria will depend on the choice of the kernel function.

Another fundamental issue of interest is to check whether the covariates really have measurement errors or not. Often, applied researchers carry out their work with the assumption on not having measurement errors, and it leads to an erroneous result. To overcome these problems, there are a few research articles investigating whether the covariates have measurement errors or not, and the methodologies adopted in those papers are bit cumbersome (see, e.g., [29]). However, one can rather think of an ad-hoc procedure to check for the presence of measurement errors in the covariates. For instance, in the absence of measurement errors in the covariates, the covariates and the regression error should be independent random variables, and in the presence of measurement errors in the covariates, the covariates and the random error are expected to be dependent random variables. Therefore, the test for independence between the covariates and the regression error is likely to give us an insight into whether the measurement error is present in the covariates or not. For details about the test for independence between the regression error and the covariates, readers can refer to [30] and a few references therein.

ACKNOWLEDGEMENT

The first author and the second author are partially supported by their MATRICS grant of SERB (Files no: MTR/2019/000033 and MTR/2019/000039, respectively), Government of India.

REFERENCES

1. Cheng, C.-L. and VanNess, J. W. (1999). Statistical regression with measurement errors. Arnold, London.
2. Klepper, Steven and Leamer, Edward E. (1984). Consistent sets of estimates for regressions with errors in all variables. Econometrica, **52**(1), 163–183.
3. Liu, M. and Li, Z. (2011). Estimation in linear mixed-effects model with errors in covariate. *International Conference on Multimedia Technology, Hangzhou*, 2458–2461.
4. Bao, X., Xiong, Z., Sheng, S., Dai, Y., Bao, S. and Liu, J. (2017). Barometer measurement error modeling and correction for UAH altitude tracking. *29th Chinese Control And Decision Conference (CCDC), Chongqing* 3166–3171.
5. Watson, R. M., Gross, J. N., Taylor, C. N. and Leishman, R. C. (2019). Enabling robust state estimation through measurement error covariance adaptation. IEEE Transactions on Aerospace and Electronic Systems.
6. Fan, J. and Troung, Y. K. (1993). Nonparametric regression with errors in variables. Annals of Statistics, **21**, 1900–1925.
7. Carroll, R. J., Maca, J. D. and Ruppert, D. (1999). Nonparametric regression in the presence of measurement error. Biometrika, **86**(3), 541–554.
8. Sørensen, Ø., Frigessi, A. and Thoresen, M. (2015). Measurement error in Lasso: Impact and likelihood bias correction. Statistica Sinica, **25**, 809–829.
9. Abarin, T. and Wang, L. (2012). Instrumental variable approach to covariate measurement error in generalized linear models. Annals of the Institute of Statistical Mathematics, **64**, 475–493.
10. Zhang, J., Zhu, L. and Zhu, L. (2014). Surrogate dimension reduction in measurement error regressions. Statistica Sinica, **24**, 1341–1363.
11. Huang, Y.-H., Wen, C-C. and Hsu, Y-H. (2015). The extensively corrected score for measurement error models. Scandinavian Journal of Statistics, **42**, 911–924.
12. Gyorfi, L., Morvai, G. and Vajda, I. (2000). Information-theoretic methods in testing the goodness of fit. *IEEE International Symposium on Information Theory (Cat. No.00CH37060), Sorrento, Italy.*
13. Song, K-S. (2002). Goodness-of-fit tests based on Kullback-Leibler discrimination information. IEEE Transactions on Information Theory, **48**(5), 1103–1117.
14. Unnikrishnan, J., Meyn, S. and Veeravalli, V. V. (2010). On thresholds for robust goodness-of-fit tests. *IEEE Information Theory Workshop, Dublin*, 1–4.
15. Kundargi, N., Liu, Y. and Tewfik, A. (2013). A framework for inference using goodness of fit tests based on ensemble of phi-divergences. IEEE Transactions on Signal Processing, **61**(4), 945–955.
16. Cheng, C.-L., Shalabh, Garg, G. (2016). Goodness of fit in restricted measurement error models. Journal of Multivariate Analysis, **145**, 101–116.
17. Anderson, T. W. (2003). An introduction to multivariate statistical analysis. John Wiley & Sons.

18. Shalabh (1998). Improved estimation in measurement error models through Stein-rule procedure. Journal of Multivariate Analysis, **67**, 35–48.
19. Shalabh (2003). Consistent estimation of coefficients in measurement error models under non-normality. Journal of Multivariate Analysis, **86**(2), 227–241.
20. Cheng, C.-L., Shalabh, Garg, G. (2014). Coefficient of determination for multiple measurement error models. Journal of Multivariate Analysis, **126**, 137–152.
21. Nadaraya, E. A. (1964). On estimating regression. Theory of Probability and its Applications, **9**, 141–142.
22. Watson, G. S. (1964). Smooth regression analysis. Sankhy A: The Indian Journal of Statistics, Series A, **26**, 359–372.
23. Priestley, M. E. and Chao, M. T. (1972). Non-parametric function fitting. Journal of The Royal Statistical Society, Series B, **34**, 385–392.
24. Silverman, B. W. (1986). Density estimation for statistics and data analysis. Chapman & Hall, London, New York.
25. Efron, B. and Gong, G. (1983). A leisurely look at the bootstrap, the jackknife, and cross-validation. The American Statistician, **37**, 36–48.
26. Battese, G. E., Fuller, W. A. and Hickman, R. D. (1976). Estimation of response variances from interview-reinterview surveys. Journal of the Indian Society of Agricultural Statistics, **28**, 1–14.
27. Fuller, W. A. (1987). Measurement error models. John Wiley & Sons.
28. Kulcsar, E. (2009). Multiple regression analysis of main economic indicators in tourism. Journal of Tourism – Studies and Research in Tourism, 8(8), 59–64.
29. Testing for the presence of measurement error in Stata. The Stata Journal, **20**, 382–404.
30. Dhar, S. S., Bergsma, W. and Dassios, A. (2018). Testing independence of covariates and errors in nonparametric regression. Scandinavian Journal of Statistics, **45**, 421–443.
31. Stuart, A and Ord, K. (1994). Kendall's advanced theory of statistics. Arnold, London.
32. Elandt-Johnson, R. C. and Johnson, N. L. (1980). Survival Models and Data Analysis. John Wiley & Sons.

APPENDIX

Lemma 1: *Under (A1)–(A6), for any* $\mathbf{z} = (z_1, \ldots, z_p) \in A$, $\hat{m}_n^{NW}(\mathbf{z}) - m(\mathbf{z}) -$ $B(\mathbf{z}) \xrightarrow{p} 0$ *as* $n \to \infty$, *where* $B(\mathbf{z}) = m(\mathbf{z}) - \dfrac{b_n^p m(\mathbf{z})\left\{\sum_{i=1}^n f(\mathbf{z} - \mathbf{x}_i)\right\}^2}{\sum_{i=1}^n f_\eta(\mathbf{z} - \mathbf{x}_i)\int k^2(\mathbf{u})d\mathbf{u}}$.

Proof of Lemma 1: *For* $\mathbf{z} \in A$, *the Nadarya-Watson estimator is defined as*

$$
\hat{m}_n^{NW}(\mathbf{z}) = \frac{\sum_{i=1}^n y_i^* K\left(\frac{\mathbf{z} - x_i^*}{b_n}\right)}{\sum_{i=1}^n K\left(\frac{\mathbf{z} - x_i^*}{b_n}\right)} = \frac{\frac{1}{nb_n^p}\sum_{i=1}^n y_i^* K\left(\frac{\mathbf{z} - x_i^*}{b_n}\right)}{\frac{1}{nb_n^p}\sum_{i=1}^n K\left(\frac{\mathbf{z} - x_i^*}{b_n}\right)} := \frac{N_n(\mathbf{z})}{D_n(\mathbf{z})}.
$$

$$
= \frac{\sum_{i=1}^n m(x_i^*) K\left(\frac{x - x_i^*}{b_n}\right) + \sum_{i=1}^n \varepsilon_i K\left(\frac{x - x_i^*}{b_n}\right)}{\sum_{i=1}^n K\left(\frac{x - x_i^*}{b_n}\right)}
$$

$$
= \frac{\sum_{i=1}^n m(x_i + \eta_i) K\left(\frac{x - x_i - \eta_i}{b_n}\right) + \sum_{i=1}^n \varepsilon_i K\left(\frac{x - x_i - \eta_i}{b_n}\right)}{\sum_{i=1}^n K\left(\frac{x - x_i^*}{b_n}\right)}.
$$

It now follows from [31] and [32] that as $n \to \infty$,

$$
E\left(\frac{N_n(\mathbf{z})}{D_n(\mathbf{z})}\right) - \left[\frac{E(N_n(\mathbf{z}))}{E(D_n(\mathbf{z}))} - \frac{\mathrm{Cov}(N_n(\mathbf{z}), D_n(\mathbf{z}))}{E(D_n(\mathbf{z}))^2}\right] \to 0 \tag{1.10}
$$

and

$$
V\left(\frac{N_n(\mathbf{z})}{D_n(\mathbf{z})}\right) - \frac{E(N_n(\mathbf{z}))^2}{E(D_n(\mathbf{z}))^2}\left[\frac{V(N_n(\mathbf{z}))}{E(N_n(\mathbf{z}))^2} + \frac{V(D_n(\mathbf{z}))}{E(D_n(\mathbf{z}))^2}\right.
$$
$$
\left. - \frac{2\mathrm{Cov}(N_n(\mathbf{z}), D_n(\mathbf{z}))}{E(N_n(\mathbf{z}))E(D_n(\mathbf{z}))}\right] \to 0. \tag{1.11}
$$

Here, after solving, we get

$$
E(N_n(\mathbf{z})) - \frac{1}{n}\sum_{i=1}^n m(\mathbf{z})f_\eta(\mathbf{z} - x_i) \to 0 \tag{1.12}
$$

and

$$
E(D_n(\mathbf{z})) - \left[\frac{1}{n}\sum_{i=1}^n f_\eta(\mathbf{z} - x_i)\right] \to 0. \tag{1.13}
$$

Arguing in a similar way, one can also establish that

$$E(D_n(\mathbf{z}))^2 - \left[\frac{1}{n^2 h_n^p} \sum_{i=1}^{n} f_\eta(\mathbf{z} - x_i) \right] \int k^2(\mathbf{u}) d\mathbf{u} \to 0. \tag{1.14}$$

and

$$E(N_n(\mathbf{z}) D_n(\mathbf{z})) - \left[\frac{1}{n^2 h_n^p} \sum_{i=1}^{n} m(\mathbf{z}) f_\eta(\mathbf{z} - x_i) \right] \int k^2(\mathbf{u}) d\mathbf{u} \to 0. \tag{1.15}$$

Therefore substituting (6.3), (6.4), (6.5), and (6.6), into approximation formula (6.1), as $n \to \infty$, and $h_n \to 0$ in such a way that $nh_n^p \to \infty$, one can conclude that

$$E(m_{n,\eta}^{NW}(\mathbf{z}) - m(\mathbf{z}) - B(\mathbf{z})) \to 0, \ where \ B(\mathbf{z}) = m(\mathbf{z})$$

$$- \frac{h_n^p m_\eta(\mathbf{z}) \left\{ \sum_{i=1}^{n} f_\eta(\mathbf{z} - \mathbf{x}_i) \right\}^2}{\sum_{i=1}^{n} f_\eta(\mathbf{z} - \mathbf{x}_i) \int k^2(\mathbf{u}) d\mathbf{u}}.$$

Following similar calculations, we can show that variance($m_n(\mathbf{z})$) $\to 0$ as $n \to \infty$, and hence, the proof is complete.

Lemma 2: *Under (A1)–(A6), for any $\mathbf{z} \in A$, $\hat{m}_n^{PC}(\mathbf{z}) \overset{p}{\to} m(\mathbf{z}) \int_A f_\eta(\mathbf{z} - \mathbf{y}) d\mathbf{y}$.*

Proof of Lemma 2: *For $\mathbf{z} \in A$, the Pristley-Chao estimator is defined as*

$$
\begin{aligned}
m_n^{PC}(\mathbf{z}) &= \frac{1}{n h_n^p} \sum_{i=1}^{n} y_i^* K\left(\frac{\mathbf{z} - x_i^\star}{h_n} \right) \\
&= \frac{1}{n h_n^p} \sum_{i=1}^{n} m(x_i^\star) K\left(\frac{\mathbf{z} - x_i^\star}{h_n} \right) + \frac{1}{n h_n^p} \sum_{i=1}^{n} \varepsilon_i K\left(\frac{\mathbf{z} - x_i^\star}{h_n} \right) \\
&= \frac{1}{n h_n^p} \sum_{i=1}^{n} m(x_i^\star - \eta_i) K\left(\frac{\mathbf{z} - x_i^\star}{h_n} \right) + \frac{1}{n h_n^p} \sum_{i=1}^{n} \varepsilon_i K\left(\frac{\mathbf{z} - x_i^\star}{h_n} \right) \\
&= \frac{1}{n h_n^p} \sum_{i=1}^{n} \left[m(xi^*) - \nabla m(xi^*) \eta_i + \frac{1}{2} \eta_i' \nabla^2 m(x_i^*) \eta_i \right] K\left(\frac{\mathbf{z} - x_i^\star}{h_n} \right) \\
&\quad + \frac{1}{n h_n^p} \sum_{i=1}^{n} \varepsilon_i K\left(\frac{\mathbf{z} - x_i^\star}{h_n} \right).
\end{aligned}
$$

This implies that

$$E(m_n^{PC}(\mathbf{z})) = \frac{1}{nb_n^p} \sum_{i=1}^{n} \int [m(x_i^*)] K\left(\frac{\mathbf{z} - x_i - \eta}{b_n}\right) f_\eta d\eta$$
$$+ \frac{1}{nb_n^p} \sum_{i=1}^{n} \int \int \varepsilon K\left(\frac{\mathbf{z} - x_i - \eta}{b_n}\right) f_\varepsilon f_\eta d\varepsilon d\eta.$$

Using the similar arguments as in the proof of Lemma 1 that $E(m_n^{PC}) \to m(\mathbf{z}) \int_A f_\eta(\mathbf{z} - y) dy$ as $n \to \infty$, and using the similar calculations, we can show that variance $(m_n)(\mathbf{z}) \to 0$ as $n \to \infty$. Hence, the proof is complete.

Lemma 3: *Under (A1)–(A6),*

$$\sqrt{nb_n^p}\left(\hat{m}_n^{NW}(\mathbf{z}) - m(\mathbf{z}) - B(\mathbf{z})\right) \overset{d}{\longrightarrow} N\left(\frac{1}{\int_A f_\eta(\mathbf{z} - y) dy}, \frac{\sigma_\varepsilon^2 \int K(u)^2 du}{\int_A f_\eta(\mathbf{z} - y) dy}\right),$$

where

$$B(\mathbf{z}) = -\nabla m(\mathbf{z}) \frac{\int_{[a,b]^p} (\mathbf{z} - y) f_\eta(\mathbf{z} - y) dy}{\int_{[a,b]^p} f_\eta(\mathbf{z} - y) dy}$$
$$+ \frac{1}{2} \frac{\int_{[a,b]^p} (\mathbf{z} - y)' \nabla^2 m(\mathbf{z})(\mathbf{z} - y) f_\eta(\mathbf{z} - y) dy}{\int_{[a,b]^p} f_\eta(\mathbf{z} - y) dy}$$

and

$$b(\mathbf{z}) = \frac{1}{2} \int \int_{[a,b]^p} (\mathbf{z} - y)' \nabla^2 m(\mathbf{z}) uu' (\nabla f_\eta(\mathbf{z} - y))' K(u) du dy$$
$$- \frac{1}{2} \int \int_{[a,b]^p} u' \nabla^2 m(\mathbf{z})(\mathbf{z} - y) \nabla f_\eta(\mathbf{z} - y) u K(u) du dy.$$

Proof of Lemma 3: *To prove the lemma, recall the model again:*

$$y_i^* = m(x_i^*) + \varepsilon_i = m(\mathbf{z}) + [m(x_i^*) - m(\mathbf{z})] + \varepsilon_i.$$

Therefore, note that

$$\frac{1}{nb_n^p} \sum_{i=1}^{n} K\left(\frac{\mathbf{z} - x_i^*}{b_n}\right) y_i^*$$
$$= \frac{1}{nb_n^p} \sum_{i=1}^{n} K\left(\frac{\mathbf{z} - x_i^*}{b_n}\right) m(\mathbf{z}) + \frac{1}{nb_n^p} \sum_{i=1}^{n} K\left(\frac{\mathbf{z} - x_i^*}{b_n}\right) [m(x_i^*) - m(\mathbf{z})]$$
$$+ \frac{1}{nb_n^p} \sum_{i=1}^{n} K\left(\frac{\mathbf{z} - x_i^*}{b_n}\right) \varepsilon_i.$$

It follows that,

$$m_{n,\eta}^{NW}(\mathbf{z}) = \frac{\frac{1}{nb_n^p}\Sigma_{i=1}^n K\left(\frac{z-x_i^*}{b_n}\right)y_i^\star}{\frac{1}{nb_n^p}\Sigma_{i=1}^n K\left(\frac{z-x_i^*}{b_n}\right)}$$

$$= m(\mathbf{z}) + \frac{\frac{1}{nb_n^p}\Sigma_{i=1}^n K\left(\frac{z-x_i^*}{b_n}\right)[m(x_i^*)-m(\mathbf{z})]}{\frac{1}{nb_n^p}\Sigma_{i=1}^n K\left(\frac{z-x_i^*}{b_n}\right)} + \frac{\frac{1}{nb_n^p}\Sigma_{i=1}^n K\left(\frac{z-x_i^*}{b_n}\right)\varepsilon_i}{\frac{1}{nb_n^p}\Sigma_{i=1}^n K\left(\frac{z-x_i^*}{b_n}\right)}$$

$$:= m(\mathbf{z}) + \frac{m_1(\mathbf{z})}{\frac{1}{nb_n^p}\Sigma_{i=1}^n K\left(\frac{z-x_i^*}{b_n}\right)} + \frac{m_2(\mathbf{z})}{\frac{1}{nb_n^p}\Sigma_{i=1}^n K\left(\frac{z-x_i^*}{b_n}\right)}.$$

We now analyse the asymptotic convergence of $m_1(\mathbf{z})$ and $m_2(\mathbf{z})$.

Let us first consider $m_2(\mathbf{z})$. Since, $E(\varepsilon_i|x_i^*) = 0$ it follows that $E\left(K\left(\frac{z-x_i^*}{b_n}\right)\varepsilon_i\right) = 0.$ *Now,*

$$E\left(\left(\frac{1}{nb_n^p}K\left(\frac{z-x_i^*}{b_n}\right)\varepsilon_i\right)^2\right)$$
$$= \frac{\sigma_\varepsilon^2}{n^2 b_n^p}\int K^2(u)\left[f_\eta(\mathbf{z}-x_i) + b_n\nabla f_\eta(x-\mathbf{z}_i)u\right]du + o(b_n)$$

Hence,

$$V\left(\frac{1}{nb_n^p}K\left(\frac{\mathbf{z}-x_i^*}{b_n}\right)\varepsilon_i\right) - \frac{\sigma_\varepsilon^2}{n^2 b_n^p}f_\eta(\mathbf{z}-x_i)\int K^2(u)\,du \to 0,$$

since first and second derivatives of f_η are bounded and $b_n \to 0$ as $n \to \infty$.

Now, let $s_n^2 = \frac{\sigma_\varepsilon^2}{n^2 b_n^p}\Sigma_{i=1}^n f_\eta(\mathbf{z}-x_i)\int K^2(u)\,du.$

Next, we have to show that $\frac{1}{nb_n^p}K\left(\frac{z-x_i^*}{b_n}\right)\varepsilon_i$ satisfies Lindeberg's condition. Therefore,

$$\lim_{n\to\infty}\frac{1}{s_n^2}\Sigma_{i=1}^n E\left[\left\{\frac{1}{nb_n^p}K\left(\frac{z-x_i^*}{b_n}\right)\varepsilon_i\right\}^2 \cdot \mathbf{1}_{\left\{\left|\frac{1}{nb_n^p}K\left(\frac{z-x_i^*}{b_n}\right)\varepsilon_i\right|>ts_n\right\}}\right] \text{ for all } t>0,$$

$$= \lim_{n\to\infty}\frac{1}{s_n^2}\Sigma_{i=1}^n E\left[\left\{\frac{1}{nb_n^p}K\left(\frac{z-x_i^*}{b_n}\right)\varepsilon_i\right\}^2 \mathbf{1}_{\left\{\left|\frac{1}{nb_n^p}K\left(\frac{z-x_i^*}{b_n}\right)\varepsilon_i\right|>ts_n\right\}}\right]$$

$$\times P\left(\left|\frac{1}{nb_n^p}K\left(\frac{z-x_i^*}{b_n}\right)\varepsilon_i\right| > ts_n\right)$$

$$= 0$$

where $\mathbf{1}_{\{.\}}$ *is the indicator function.*

Thus according to Lindeberg's CLT, we have

$$\sqrt{nh_n^p}\, m_2(\mathbf{z}) \xrightarrow{d} N\left(0,\, \sigma_\varepsilon^2 \int K(u)^2 du \int_A f_\eta(\mathbf{z}-y)\,dy\right).$$

Now consider $m_1(\mathbf{z})$. *Note that*

$$E(m_1(\mathbf{z})) = \frac{1}{n}\sum_{i=1}^n \int K(u)[m(\mathbf{z}+uh_n) - m(\mathbf{z})]f_\eta(\mathbf{z}-x_i+uh_n)\,du.$$

Arguing similarly as the proofs of Lemma 1, *we have* $E(m_1(\mathbf{z})) - \frac{1}{n}\sum_{i=1}^n f_\eta(\mathbf{z}-x_i) \to 0$ *as* $n \to \infty$. *Moreover, using the similar calculation shows, we can show that* nh_n^p *variance* $(m_1(\mathbf{z})) \to 0$ *as* $n \to \infty$. *Hence,* $\sqrt{nh_n^p}\,\{m_1(\mathbf{z}) - \frac{1}{n}\sum_{i=1}^n f_\eta(\mathbf{z}-x_i)\} \xrightarrow{p} 0$ *as* $n \to \infty$.

Hence,

$$\sqrt{nh_n^p}\,(m_n^{NW}(\mathbf{z}) - m(\mathbf{z}) - B(\mathbf{z})) \xrightarrow{d} N\left(\frac{1}{\int_A f_\eta(\mathbf{z}-y)\,dy},\, \frac{\sigma_\varepsilon^2 \int K(u)^2 du}{\int_A f_\eta(\mathbf{z}-y)\,dy}\right).$$

Lemma 4: *Under (A1)–(A6),*

$$\sqrt{nh_n^p}\left(\hat{m}_n^{PC}(\mathbf{z}) - m(\mathbf{z})\int_A f_\eta(\mathbf{z}-y)\,dy - \tilde{B}(\mathbf{z})\right)$$
$$\xrightarrow{d} N\left(0,\, \sigma_\varepsilon^2 \int K(u)^2 du \int_A f_\eta(\mathbf{z}-y)\,dy\right),$$

where

$$\tilde{B}(\mathbf{z}) = \int_A f_\eta(\mathbf{z}-\mathbf{y})\,d\mathbf{y},$$
$$\tilde{B}(\mathbf{z}) = -\nabla m(\mathbf{z})\int_{[a,b]^p}(\mathbf{z}-y)f_\eta(\mathbf{z}-y)\,dy$$
$$+ \frac{1}{2}\int_{[a,b]^p}(\mathbf{z}-y)'\nabla^2 m(\mathbf{z})(\mathbf{z}-y)f_\eta(\mathbf{z}-y)\,dy$$

and

$$b(\mathbf{z}) = \frac{1}{2}\int\int_{[a,b]^p}(\mathbf{z}-y)'\nabla^2 m(\mathbf{z})uu'(\nabla f_\eta(\mathbf{z}-y))'K(u)\,du\,dy$$
$$- \frac{1}{2}\int\int_{[a,b]^p} u'\nabla^2 m(\mathbf{z})(\mathbf{z}-y)\nabla f_\eta(\mathbf{z}-y)uK(u)\,du\,dy.$$

Proof of Lemma 4: *To prove the lemma, recall the model again:*

$$y_i^* = m(x_i^*) + \varepsilon_i = m(\mathbf{z}) + [m(x_i^*) - m(\mathbf{z})] + \varepsilon_i.$$

Therefore, note that

$$
\begin{aligned}
m_n^{PC}(\mathbf{z}) &= \frac{1}{nh_n^p} \sum_{i=1}^{n} K\left(\frac{\mathbf{z} - x_i^*}{h_n}\right) y_i^* \\
&= \frac{1}{nh_n^p} \sum_{i=1}^{n} K\left(\frac{\mathbf{z} - x_i^*}{h_n}\right) m(\mathbf{z}) + \frac{1}{nh_n^p} \sum_{i=1}^{n} K\left(\frac{\mathbf{z} - x_i^*}{h_n}\right)[m(x_i^*) - m_\eta(\mathbf{z})] \\
&\quad + \frac{1}{nh_n^p} \sum_{i=1}^{n} K\left(\frac{\mathbf{z} - x_i^*}{h_n}\right) \varepsilon_i \\
&= \frac{1}{nh_n^p} \sum_{i=1}^{n} K\left(\frac{\mathbf{z} - x_i^*}{h_n}\right) m(\mathbf{z}) + m_1(\mathbf{z}) + m_2(\mathbf{z}),
\end{aligned}
$$

where

$$m_1(\mathbf{z}) = \frac{1}{nh_n^p} \sum_{i=1}^{n} K\left(\frac{\mathbf{z} - x_i^*}{h_n}\right)[m(x_i^*) - m_\eta(\mathbf{z})],$$

and

$$m_2(\mathbf{z}) = \frac{1}{nh_n^p} \sum_{i=1}^{n} K\left(\frac{\mathbf{z} - x_i^*}{h_n}\right) \varepsilon_i.$$

As we did in Lemma 3, *we here also analyse the asymptotic convergence of* $m_1(\mathbf{z})$ *and* $m_2(\mathbf{z})$.

First, we consider $m_2(\mathbf{z})$. As $E(\varepsilon_i | x_i^*) = 0$, it leads to $E\left(K\left(\frac{\mathbf{z} - x_i^*}{h_n}\right)\varepsilon_i\right) = 0$. Now,

$$
\begin{aligned}
&E\left(\left(\frac{1}{nh_n^p} K\left(\frac{\mathbf{z} - x_i^*}{h_n}\right)\varepsilon_i\right)^2\right) \\
&= \frac{1}{n^2 h_n^{2p}} \iint \varepsilon^2 K^2\left(\frac{\mathbf{z} - x_i - \eta}{h_n}\right) f_\varepsilon f_\eta \, d\varepsilon d\eta \\
&= \frac{\sigma_\varepsilon^2}{n^2 h_n^{2p}} \int K^2\left(\frac{\mathbf{z} - x_i - \eta}{h_n}\right) f_\eta \, d\eta.
\end{aligned}
$$

Transformation $\frac{\mathbf{z} - x_i - \eta}{h_n} = -u$ *leads to*

$$E\left(\left(\frac{1}{nh_n^p} K\left(\frac{\mathbf{z} - x_i^*}{h_n}\right)\varepsilon_i\right)^2\right) = \frac{\sigma_\varepsilon^2}{n^2 h_n^p} \int K(u)^2 f_\eta(\mathbf{z} - x_i + u h_n) \, du.$$

Further application of Taylor series expansion, and as second derivative of f_η is bounded, it leads to

$$= \frac{\sigma_\varepsilon^2}{n^2 h_n^p} \int K^2(u)[f_\eta(z - x_i) + h_n \nabla f_\eta(x - z_i)u]\,du + o(h_n)$$

Hence,

$$V\left(\frac{1}{nh_n^p}K\left(\frac{z - x_i^*}{h_n}\right)\varepsilon_i\right) - \frac{\sigma_\varepsilon^2}{n^2 h_n^p}f_\eta(z - x_i)\int K^2(u)\,du \to 0,$$

since first and second derivatives of f_η are bounded, and $h_n \to 0$ as $n \to \infty$.

Now, let $s_n^2 = \frac{\sigma_\varepsilon^2}{n^2 h_n^p}\sum_{i=1}^n f_\eta(z - x_i)\int K^2(u)\,du$. Next, one needs to show that $\frac{1}{nh_n^p}K\left(\frac{z - x_i^*}{h_n}\right)\varepsilon_i$ satisfies Lindeberg's condition. Therefore,

$$\lim_{n \to \infty} \frac{1}{s_n^2}\sum_{i=1}^n E\left[\left\{\frac{1}{nh_n^p}K\left(\frac{z - x_i^*}{h_n}\right)\varepsilon_i\right\}^2 \cdot \mathbf{1}_{\left\{\left|\frac{1}{nh_n^p}K\left(\frac{z - x_i^*}{h_n}\right)\varepsilon_i\right| > ts_n\right\}}\right] \quad for \ all \ t > 0.$$

$$= \lim_{n \to \infty} \frac{1}{s_n^2}\sum_{i=1}^n E\left[\left\{\frac{1}{nh_n^p}K\left(\frac{z - x_i^*}{h_n}\right)\varepsilon_i\right\}^2 \middle| \mathbf{1}_{\left\{\left|\frac{1}{nh_n^p}K\left(\frac{z - x_i^*}{h_n}\right)\varepsilon_i\right| > ts_n\right\}}\right]$$

$$\times P\left(\left|\frac{1}{nh_n^p}K\left(\frac{z - x_i^*}{h_n}\right)\varepsilon_i\right| > ts_n\right)$$

$$= 0.$$

Thus according to Lindeberg's CLT, we have

$$\sqrt{nh_n^p}\, m_2(z) \xrightarrow{d} N\left(0, \sigma_\varepsilon^2 \int K(u)^2 du \int_A f_\eta(z - y)\,dy\right).$$

Now consider $m_1(z)$ and derive

$$m_1(z) = \frac{1}{nh_n^p}\sum_{i=1}^n K\left(\frac{z - x_i^*}{h_n}\right)[m(x_i^*) - m(z)]$$

$$= \frac{1}{nh_n^p}\sum_{i=1}^n K\left(\frac{z - x_i - \eta_i}{h_n}\right)[m(x_i + \eta_i) - m(z)]$$

$$E(m_1(z)) = \frac{1}{nh_n^p}\sum_{i=1}^n \int K\left(\frac{z - x_i - \eta_i}{h_n}\right)[m(x_i + \eta_i) - m(z)]f_\eta\,d\eta.$$

The transformation $\frac{z - x_i - \eta}{b_n} = -u$, leads us to

$$E(m_1(\mathbf{z})) = \frac{1}{n} \sum_{i=1}^{n} \int K(u)[m(\mathbf{z} + ub_n) - m(\mathbf{z})]f_\eta(\mathbf{z} - x_i + ub_n)\,du.$$

Moreover, further application of Taylor series expansion, and in view of the fact that f_η has a bounded second derivative, we have $E(m_1(\mathbf{z})) \to \int_A f_\eta(\mathbf{z} - \mathbf{y})\,d\mathbf{y}$ as $n \to \infty$ following the same arguments in Lemma 2. Furthermore, similar arguments with similar algebraic nh_n^p variance$(m_1(\mathbf{z})) \to 0$ as $n \to \infty$. Hence, the proof is complete.

Lemma 5: $\widehat{\sigma_{\varepsilon,n}^2}$ *is a consistent estimator of σ_ε^2, where* $\widehat{\sigma_{\varepsilon,n}^2} = \frac{1}{n-1}\sum_{i=1}^{n}$ $\left(y_i^* - \hat{m}_n(\mathbf{x}_i^*)\right)^2$.

Proof of Lemma 5: *The estimator is defined as* $\widehat{\sigma_{\varepsilon,n}^2} = \frac{1}{n-1}\sum_{i=1}^{n}$ $\left(y_i^* - m_{n,\eta}(\mathbf{z})\right)^2$, *where $m_{n,\eta}(\mathbf{z})$ is the estimator of $m(\mathbf{z})$ (see the Lemmas 1 and 2).*
   Observe that

$$\widehat{\sigma_{\varepsilon,n}^2} = \frac{1}{n-1} \sum_{i=1}^{n} \left(y_i^* - m_n(\mathbf{x}_i^*)\right)^2 = \frac{1}{n-1} \sum_{i=1}^{n} [\varepsilon_i - (m_n(\mathbf{x}_i^*) - m(\mathbf{x}_i^*))]^2.$$

This implies

$$
\begin{aligned}
E\left(\widehat{\sigma_{\varepsilon,n}^2}\right) &= \tfrac{1}{n-1}\sum_{i=1}^{n} E\,[\varepsilon_i - (m_n(\mathbf{x}_i^*) - m(\mathbf{x}_i^*))]^2 \\
&= \tfrac{1}{n-1}\sum_{i=1}^{n} E\,[\varepsilon_i^2 + (m_n(\mathbf{x}_i^*) - m(\mathbf{x}_i^*))^2 + 2\varepsilon_i(m_n(\mathbf{x}_i^*) - m(\mathbf{x}_i^*))] \\
&= \tfrac{1}{n-1}\sum_{i=1}^{n} [\sigma_\varepsilon^2 + E(m_n(\mathbf{x}_i^*) - m(\mathbf{x}_i^*))^2]
\end{aligned}
$$

(since $E(\varepsilon) = 0$, $E(\varepsilon^2) = \sigma_\varepsilon^2$, and ε and η are independent)

$$= \frac{n}{n-1}\sigma_\varepsilon^2 + \frac{1}{n-1}\sum_{i=1}^{n} E(m_n(\mathbf{x}_i^*) - m(\mathbf{x}_i^*))^2.$$

It follows from the assertions Lemma 3 and Lemma 4 along with an application of dominated convergence theorem that $E(m_n(\mathbf{x}_i^*) - m(\mathbf{x}_i^*))^2 \to 0$ as $n \to \infty$.

Hence, $E(\widehat{\sigma^2_{\varepsilon,n}}) \to \sigma^2_\varepsilon$ as $n \to \infty$. Similar argument and calculations further leads to variance $(\widehat{\sigma^2_{\varepsilon,n}}) \to 0$ as $n \to \infty$. Hence, the proof is complete.

Proof of Theorem 1: To prove this theorem, let us denote $R^2_{n,\mathrm{NW}}(\mathbf{z}) = \frac{(m_n^{NW}(\mathbf{z}))^2}{\sigma^2_{\varepsilon,n} + (m_n^{NW}(\mathbf{z}))^2}$ and $\rho^2(\mathbf{z}) = \frac{m(\mathbf{z})}{\sigma^2_\varepsilon + m(\mathbf{z})}$. Now, using Lemmas 1, 3, and 5, we have

$$\sqrt{nb_n^p}\left(R^2_{n,\mathrm{NW}}(\mathbf{z}) - \rho^2(\mathbf{z})\right)$$

$$= \sqrt{nb_n^p}\,\sigma^2_\varepsilon \left[\frac{(m_n^{NW}(\mathbf{z}))^2 - (m(\mathbf{z}) + B(\mathbf{z}))^2}{(\sigma^2_\varepsilon + (m_n^{NW}(\mathbf{z}))^2)(\sigma^2_\varepsilon + m^2(\mathbf{z}))}\right]$$

$$\sqrt{nb_n^p}\,\sigma^2_\varepsilon \left[\frac{2m(\mathbf{z})B(\mathbf{z}) + B^2(\mathbf{z})}{(\sigma^2_\varepsilon + (m_n^{NW}(\mathbf{z}))^2)(\sigma^2_\varepsilon + m^2(\mathbf{z}))}\right].$$

As $m_n^{NW}(\mathbf{z}) \xrightarrow{p} m(\mathbf{z}) + B(\mathbf{z})$ (see Lemma 1), we have

$$\sqrt{nb_n^p}\left(R^2_{n,\mathrm{NW}}(\mathbf{z}) - \rho^2(\mathbf{z}) - P(\mathbf{z})\right) \xrightarrow{d} \frac{\sigma^2_\varepsilon T_1(\mathbf{z})}{(\sigma^2_\varepsilon + (m(\mathbf{z}) + B(\mathbf{z}))^2)(\sigma^2_\varepsilon + m^2(\mathbf{z}))},$$

where $P(\mathbf{z}) = \sigma^2_\varepsilon\left[\frac{2m(\mathbf{z})B(\mathbf{z}) + B^2(\mathbf{z})}{(\sigma^2_\varepsilon + (m(\mathbf{z}) + B(\mathbf{z}))^2)(\sigma^2_\varepsilon + m^2(\mathbf{z}))}\right]$, and $T_1(\mathbf{z})$ is a random variable associated with normal distribution with mean $= \frac{2(m(\mathbf{z}) + B(\mathbf{z}))}{\int_A f_\eta(\mathbf{z} - y)dy}$ and variance $= \frac{4(m_\eta(\mathbf{z}) + B(\mathbf{z}))^2\sigma^2_\varepsilon \int k^2(u)du}{\int_A f_\eta(\mathbf{z} - y)dy}$. Finally, the proof of this theorem follows by direct application of continuous mapping theorem as the integration operator is a continuous operator. Note that in the statement of the theorem in the main manuscript, the unknown terms are the following. $a_1(\mathbf{z}) = \frac{2(m(\mathbf{z}) + B(\mathbf{z}))}{\int_A f_\eta(\mathbf{z} - y)dy}$, $b_1(\mathbf{z}) = \frac{4(m(\mathbf{z}) + B(\mathbf{z}))^2\sigma^2_\varepsilon \int k^2(u)du}{\int_A f_\eta(\mathbf{z} - y)dy}$, $P(\mathbf{z}) = \sigma^2_\varepsilon\left[\frac{2m(\mathbf{z})B(\mathbf{z}) + B^2(\mathbf{z})}{(\sigma^2_\varepsilon + (m(\mathbf{z}) + B(\mathbf{z}))^2)(\sigma^2_\varepsilon + m^2(\mathbf{z}))}\right]$ and $B(\mathbf{z}) = m(\mathbf{z}) - \frac{b_n^p m(\mathbf{z})\left\{\sum_{i=1}^n f_\eta(\mathbf{z} - \mathbf{x}_i)\right\}^2}{\sum_{i=1}^n f_\eta(\mathbf{z} - \mathbf{x}_i)\int k^2(u)du}$.

Proof of Theorem 2: To prove this theorem, let us denote $R^2_{n,\mathrm{PC}}(\mathbf{z}) = \frac{(m_n^{PC}(\mathbf{z}))^2}{\sigma^2_{\varepsilon,n} + (m_n^{PC}(\mathbf{z}))^2}$, and as we denoted in the proof of Theorem 1, $\rho^2(\mathbf{z}) = \frac{m(\mathbf{z})}{\sigma^2_\varepsilon + m(\mathbf{z})}$. Using Lemmas 2, 4, and 5, we have

$$\sqrt{nb_n^p}\left(R^2_{n,\mathrm{PC}}(\mathbf{z}) - \rho^2(\mathbf{z})\right)$$

$$= \sqrt{nb_n^p}\left[\frac{(m^{PC}(\mathbf{z}))^2}{\widehat{\sigma_{\varepsilon,n}^2} + (m_n^{PC}(\mathbf{z}))^2} - \frac{m^2(\mathbf{z})}{\sigma_\varepsilon^2 + m^2(\mathbf{z})}\right]$$

$$= \sqrt{nb_n^p}\left[\frac{(m_n^{PC}(\mathbf{z}))^2\sigma_\varepsilon^2 + (m_n^{PC}(\mathbf{z}))^2 m^2(\mathbf{z}) - m^2(\mathbf{z})\widehat{\sigma_{\varepsilon,n}^2} - m^2(\mathbf{z})(m_n^{PC}(\mathbf{z}))^2}{(\widehat{\sigma_{\varepsilon,n}^2} + (m_n^{PC}(\mathbf{z}))^2)(\sigma_\varepsilon^2 + m^2(\mathbf{z}))}\right]$$

$$\to \sqrt{nb_n^p}\,\sigma_\varepsilon^2\left[\frac{(m_n^{PC}(\mathbf{z}))^2 - m^2(\mathbf{z})}{(\sigma_\varepsilon^2 + (m_n^{PC}(\mathbf{z}))^2)(\sigma_\varepsilon^2 + m^2(\mathbf{z}))}\right] \quad (\text{as } \widehat{\sigma_{\varepsilon,n}^2} \xrightarrow{p} \sigma_\varepsilon^2 \ \ (\text{see Lemma 5}))$$

$$= \sqrt{nb_n^p}\,\sigma_\varepsilon^2\left[\frac{(m_n^{PC}(\mathbf{z}))^2 - \left(m_\eta(\mathbf{z})\int_A f_\eta(\mathbf{z}-y)\,dy\right)^2 \big/ \left(\int_A f_\eta(\mathbf{z}-y)\,dy\right)^2}{(\sigma_\varepsilon^2 + (m_n^{PC}(\mathbf{z}))^2)(\sigma_\varepsilon^2 + m^2(\mathbf{z}))}\right]$$

$$+ \sqrt{nb_n^p}\,\sigma_\varepsilon^2\left[\frac{2\left(m(\mathbf{z})\int_A f_\eta(\mathbf{z}-y)\,dy\right)\tilde{B}(\mathbf{z}) + \tilde{B}^2(\mathbf{z})}{(\sigma_\varepsilon^2 + (m_n^{PC}(\mathbf{z}))^2)(\sigma_\varepsilon^2 + m^2(\mathbf{z}))}\right]$$

$$+ \sqrt{nb_n^p}\,\sigma_\varepsilon^2\left[\frac{\left(m(\mathbf{z})\int_A f_\eta(\mathbf{z}-y)\,dy\right)^2 - \left(m(\mathbf{z})\int_A f_\eta(\mathbf{z}-y)\,dy\right)^2\big/\left(\int_A f_\eta(\mathbf{z}-y)\,dy\right)^2}{(\sigma_\varepsilon^2 + (m_n^{PC}(\mathbf{z}))^2)(\sigma_\varepsilon^2 + m^2(\mathbf{z}))}\right]$$

$$= \sqrt{nb_n^p}\,\sigma_\varepsilon^2\left[\frac{(m_n^{PC}(\mathbf{z}))^2 - \left(m(\mathbf{z})\int_A f_\eta(\mathbf{z}-y)\,dy\right)^2}{(\sigma_\varepsilon^2 + (m_n^{PC}(\mathbf{z}))^2)(\sigma_\varepsilon^2 + m^2(\mathbf{z}))}\right]$$

$$+ \sqrt{nb_n^p}\,\sigma_\varepsilon^2\left[\frac{\left(m(\mathbf{z})\int_A f_\eta(\mathbf{z}-y)\,dy\right)^2 - \left(m(\mathbf{z})\int_A f_\eta(\mathbf{z}-y)\,dy\right)^2\big/\left(\int_A f_\eta(\mathbf{z}-y)\,dy\right)^2}{(\sigma_\varepsilon^2 + (m_n^{PC}(\mathbf{z}))^2)(\sigma_\varepsilon^2 + m^2(\mathbf{z}))}\right]$$

$$= \sqrt{nh_n^p}\,\sigma_\varepsilon^2 \left[\frac{(m_n^{PC}(\mathbf{z}))^2 - \left(m(\mathbf{z})\int_A f_\eta(\mathbf{z}-y)\,dy + \tilde{B}(\mathbf{z})\right)^2}{(\sigma_\varepsilon^2 + (m_n^{PC}(\mathbf{z}))^2)(\sigma_\varepsilon^2 + m^2(\mathbf{z}))} \right]$$

$$+ \sqrt{nh_n^p}\,\sigma_\varepsilon^2 \left[\frac{2\left(m(\mathbf{z})\int_A f_\eta(\mathbf{z}-y)\,dy\right)\tilde{B}(\mathbf{z}) + \tilde{B}^2(\mathbf{z})}{(\sigma_\varepsilon^2 + (m_n^{PC}(\mathbf{z}))^2)(\sigma_\varepsilon^2 + m^2(\mathbf{z}))} \right]$$

$$+ \sqrt{nh_n^p}\,\sigma_\varepsilon^2 \left[\frac{\left(m(\mathbf{z})\int_A f_\eta(\mathbf{z}-y)\,dy\right)^2 - \left(m(\mathbf{z})\int_A f_\eta(\mathbf{z}-y)\,dy\right)^2 \Big/ \left(\int_A f_\eta(\mathbf{z}-y)\,dy\right)^2}{(\sigma_\varepsilon^2 + (m_n^{PC}(\mathbf{z}))^2)(\sigma_\varepsilon^2 + m^2(\mathbf{z}))} \right]$$

$$+ \sqrt{nh_n^p}\,\sigma_\varepsilon^2 \left[\frac{2\left(m(\mathbf{z})\int_A f_\eta(\mathbf{z}-y)\,dy\right)\tilde{B}(\mathbf{z}) + \tilde{B}^2(\mathbf{z})}{(\sigma_\varepsilon^2 + (m_n^{PC}(\mathbf{z}))^2)(\sigma_\varepsilon^2 + m^2(\mathbf{z}))} \right]$$

$$+ \sqrt{nh_n^p}\,\sigma_\varepsilon^2 \left[\frac{\left(m(\mathbf{z})\int_A f_\eta(\mathbf{z}-y)\,dy\right)^2 - \left(m(\mathbf{z})\int_A f_\eta(\mathbf{z}-y)\,dy\right)^2 \Big/ \left(\int_A f_\eta(\mathbf{z}-y)\,dy\right)^2}{(\sigma_\varepsilon^2 + (m_n^{PC}(\mathbf{z}))^2)(\sigma_\varepsilon^2 + m^2(\mathbf{z}))} \right].$$

As $m_n^{PC}(\mathbf{z}) \xrightarrow{p} m(\mathbf{z})\int_A f_\eta(\mathbf{z}-y)\,dy$ (see Lemma 2), one can conclude

$$\sqrt{nh_n^p}\,(R_n^{PC,2}(\mathbf{z}) - \rho^2(\mathbf{z}) - Q(\mathbf{z}))$$

$$\xrightarrow{d} \frac{\sigma_\varepsilon^2\,T_2(\mathbf{z})}{\left(\sigma_\varepsilon^2 + \left(m(\mathbf{z})\int_A f_\eta(\mathbf{z}-y)\,dy\right)^2\right)(\sigma_\varepsilon^2 + m^2(\mathbf{z}))},$$

where

$$Q(\mathbf{z}) = \sigma_\varepsilon^2 \left[\frac{2\left(m(\mathbf{z})\int_A f_\eta(\mathbf{z}-y)\,dy\right)\tilde{B}(\mathbf{z}) + \tilde{B}^2(\mathbf{z})}{\left(\sigma_\varepsilon^2 + \left(m(\mathbf{z})\int_A f_\eta(\mathbf{z}-y)\,dy\right)^2\right)(\sigma_\varepsilon^2 + m^2(\mathbf{z}))} \right.$$

$$\left. + \sigma_\varepsilon^2 \frac{\left(m(\mathbf{z})\int_A f_\eta(\mathbf{z}-y)\,dy\right)^2 - \frac{\left(m(\mathbf{z})\int_A f_\eta(\mathbf{z}-y)\,dy\right)^2}{\left(\int_A f_\eta(\mathbf{z}-y)\,dy\right)^2}}{\left(\sigma_\varepsilon^2 + \left(m(\mathbf{z})\int_A f_\eta(\mathbf{z}-y)\,dy\right)^2\right)(\sigma_\varepsilon^2 + m^2(\mathbf{z}))} \right]$$

and

$$T_2(\mathbf{z}) = \sqrt{nb_n^p}\left((m_n^{PC}(\mathbf{z}))^2 - \left(m(\mathbf{z})\int_A f_\eta(\mathbf{z} - y)\,dy + \tilde{B}(\mathbf{z})\right)^2\right)$$

$$\stackrel{d}{=} 2\left(m(\mathbf{z})\int_A f_\eta(\mathbf{z} - y)\,dy + \tilde{B}(\mathbf{z})\right)$$

$$\left[\sqrt{nb_n^p}\left(m_n^{PC}(\mathbf{z}) - m(\mathbf{z})\int_A f_\eta(\mathbf{z} - y)\,dy - \tilde{B}(\mathbf{z})\right)\right]$$

(using Delta method).
Substituting the assertion of Lemma 4, we have

$$\sqrt{nb_n^p}\left(R_{n,PC}^2(\mathbf{z}) - \rho^2(\mathbf{z}) - Q(\mathbf{z})\right)$$

$$\stackrel{d}{\to} \int_A \frac{\sigma_\varepsilon^2 T_2(\mathbf{z})}{\Lambda(A)\left(\sigma_\varepsilon^2 + \left(m_\eta(\mathbf{z})\int_A f_\eta(\mathbf{z} - y)\,dy + \tilde{B}(\mathbf{z})\right)^2\right)\left(\sigma_\varepsilon^2 + \{m_\eta(\mathbf{z})\}^2\right)}\,d\mathbf{z},$$

where $T_2(\mathbf{z})$ is a random variable associated with normal distribution with mean 0 and variance $4\left(m(\mathbf{z})\int_A f_\eta(\mathbf{z} - y)\,dy + \tilde{B}(\mathbf{z})\right)^2\sigma_\varepsilon^2\int K(u)^2\,du\int_A f_\eta(\mathbf{z} - y)\,dy$.
Finally, the proof of this theorem follows by direct application of continuous mapping theorem as the integration operator is a continuous operator.
Note that in the statement of the theorem in the main manuscript, the unknown terms are the following.

$$a_2(\mathbf{z}) = 0,$$
$$b_2(\mathbf{z}) = 4\left(m(\mathbf{z})\int_A f_\eta(\mathbf{z} - y)\,dy + \tilde{B}(\mathbf{z})\right)^2\sigma_\varepsilon^2\int K(u)^2\,du\int_A f_\eta(\mathbf{z} - y)\,dy,$$
$$\tilde{B}(\mathbf{z}) = \int_A f_\eta(\mathbf{z} - y)\,d\mathbf{y}$$

and

$$Q(\mathbf{z}) = \sigma_\varepsilon^2\left[\frac{m^2(\mathbf{z})\left(\left(\int_A f_\eta(\mathbf{z} - y)\,dy\right)^2 - 1\right) + 2\left(m(\mathbf{z})\int_A f_\eta(\mathbf{z} - y)\,dy\right)\tilde{B}(\mathbf{z}) + \tilde{B}^2(\mathbf{z})}{\left(\sigma_\varepsilon^2 + \left(m(\mathbf{z})\int_A f_\eta(\mathbf{z} - y)\,dy\right)^2\right)(\sigma_\varepsilon^2 + m^2(\mathbf{z}))}\right].$$

Chapter 2

Bayesian statistics with applications in cosmology

Suresh Kumar, Shivani Sharma, Rafael C. Nunes, and Divisão de Astrofísica

2.1 ELEMENTS OF BAYESIAN STATISTICS

In statistics, two classical frameworks interpret the exact nature of probability, namely the frequentist approach and the Bayesian approach. In the frequentist approach, the probability is interpreted as the relative frequency in random experiments with many trials:

$$P = \frac{\text{Number of successes}}{\text{Number of total trials}} \qquad (2.1)$$

One cannot rely on the frequentist approach in cases where it is impractical or impossible to repeat the experiment; for instance, cosmologists are unable to create multiple Universes for models. In Bayesian statistics, one deals with the measure of the "degree of belief" for a given event to occur. It is fundamentally related to our prior knowledge of an event or situation [1,2]. It depends on the two elements of sample space: data and model. The probabilities are related to our knowledge, and thus we can easily connect probability distribution to our parameters. We consider P_θ as the conditional probability of X for the given parameter θ. Therefore, in the Bayesian approach, data observed from the sample remain fixed while the parameters are unknown.

2.1.1 Bayes' theorem

Bayesian statistical tools are widely used in the area of cosmology for estimating parameters and comparing different cosmological models [3–7].

DOI: 10.1201/9781003356653-2

We define Bayes' theorem for any model M, the cosmological data D, and the set of parameters Θ as

$$P(\Theta|D, M) = \frac{\mathscr{L}(D|\Theta, M)\pi(\Theta|M)}{\mathscr{E}(D|M)}. \tag{2.2}$$

Here, $P(\Theta|D, M)$ represents the posterior distribution, $\mathscr{L}(D|\Theta, M)$ is the likelihood, $\pi(\Theta|M)$ denotes the prior probability, and the Bayesian evidence $\mathscr{E}(D|M)$ is defined via the relation

$$\mathscr{E}(D|M) = \int_M \mathscr{L}(D|\Theta, M)\pi(\Theta|M)\,d\Theta. \tag{2.3}$$

Bayesian evidence is widely used in comparing different cosmological models, while we use multivariate Gaussian likelihood for parameter estimation. For chi-squared function $\chi^2(D|\Theta, M)$, we represent the likelihood as

$$\mathscr{L}(D|\Theta, M) \propto exp\left[-\frac{\chi^2(D|\Theta, M)}{2}\right], \tag{2.4}$$

If we consider uniform prior distribution and use equation (2.2), the posterior distribution can be written as

$$P(\Theta|D, M) \propto exp\left[-\frac{\chi^2(D|\Theta, M)}{2}\right]. \tag{2.5}$$

In other words, we can say that if $\chi^2(D|\Theta, M)$ is minimum, the likelihood or the posterior probability is maximum.

2.1.2 Bayesian model selection

Bayesian evidence plays a vital role in comparing different cosmological models. For two models M_a and M_b, the ratio of their posterior probabilities is evaluated using Bayes' factor B_{ab} [8], written as

$$\frac{P(M_a|D)}{P(M_b|D)} = B_{ab}\frac{P(M_a)}{P(M_b)}, \tag{2.6}$$

and

$$B_{ab} = \frac{\mathscr{E}_a}{\mathscr{E}_b}. \tag{2.7}$$

Table 2.1   Representation of Jeffrey's scale

$|lnB_{ij}|$	Strength of evidence
[0,1)	Inconclusive/Weak
[1,3)	Definite/Positive
[3,5)	Strong
≥ 5	Very strong

The different scale ranges of the Bayes factor represent the strength of evidence, which is interpreted using Jeffery's scale [9]. Table 2.1 represents the strength of the evidence against or in favor of model M_a relative to model M_b. If $|lnB_{ab}| < 1$, there is a weak or inconclusive evidence for model M_a relative to the model M_b.

2.2 MONTE CARLO MARKOV CHAIN METHOD

The estimation of parameters and model comparison in a large parameter space becomes a little difficult and hence, faces few numerical challenges. Parameter estimation of large parameter space is done via sampling, which helps in finding the peak and shape of the joint probability distribution. On the other hand, in model comparison it is necessary to solve the large multidimensional integral. These challenges are easy in the case of low dimensional problems with fewer unknown parameters. However, as we consider the extension (interacting dark sector) of the standard ΛCDM model in this chapter, the number of parameters will increase, and hence evaluating the likelihood function and the posterior over a grid is not economical and practical. Therefore, instead of calculating the posterior distribution at each point of the parameter space, it is feasible to consider those points which are concentrated on or around the peak of the posterior distribution. In this chapter, we use the Monte Carlo Markov Chain (MCMC) method that explores the posterior in this way with the Metropolis-Hastings algorithm [10,11], and in addition to parameter inference, we compute Bayesian evidence for model selection using nested sampling technique [12,13]. We now give a quick overview of the MCMC method and nested sampling technique used in this chapter (for a detailed description refer to [14,15]).

In the MCMC technique, a sequence of points is a built-in space of parameters for evaluating the posterior distribution. *Monte Carlo* simulation is a technique that is used to generate random numbers for approximating any specific quantity or uncertainty of a system. On the other hand, *Markov Chain* is a process in which the transition of state takes place from one state to another in accordance to some probability rules. It is based on the information given at the present state, while the future states remain fixed. The chain converges to a state (often known as a stationary state) whose elements are derived from the target distribution. For our scenario, the chain converges to $P(\Theta|D, M)$, i.e., the posterior distribution. Henceforth, we can evaluate the quantities of our interest like mean, variance, etc. Thus, MCMC is a combination of both these procedures. This method is faster than the grid technique as the number of points contributing to the good estimates in MCMC is in a linear relationship with the number of parameters, and hence increasing its dimensionality. We use the set of delta functions to evaluate the target density as

$$p(\theta|D, M) \simeq \frac{1}{N} \sum_{i=1}^{N} \delta(\theta - \theta_i). \qquad (2.8)$$

Here, N denotes total number of points in the corresponding chain. Hence, the posterior mean is evaluated as

$$\langle \theta \rangle = \int \theta P(\theta, M|D) \, d\theta \simeq \frac{1}{N} \sum_{i=1}^{N} \theta_i. \qquad (2.9)$$

Therefore, for evaluating quantities like mean or variance, we can easily estimate the integral as

$$\langle f(\theta) \rangle \simeq \frac{1}{N} \sum_{i=1}^{N} f(\theta_i). \qquad (2.10)$$

For generating a new point, say θ_{i+1}, we need a criterion or technique that helps in accepting or rejecting this new point as per the model. The simplest algorithm used in this regard is the Metropolis-Hastings algorithm. In this algorithm, a random start point θ_i is chosen from a parameter space, with the un-normalized posterior $p_i = p(\theta_i|D, M)$. A candidate point θ_c is then proposed and is treated as a jump or generator of the new possible random steps, with proposal distribution $q(\theta_i, \theta_c)$ and posterior probability p_c. The probability at which a candidate point is accepted is calculated as

$$p(\text{accept}) = \min\left[1, \frac{p_c q(\theta_c, \theta_i)}{p_i q(\theta_i, \theta_c)}\right]. \qquad (2.11)$$

Mostly, the proposal distribution follows Gaussian which is similar to the target distribution. On the other hand, if we consider the symmetric proposal distribution, we observe a reversible chain, and the algorithm is called the *Metropolis algorithm*:

$$p\,(\text{accept}) \;=\; \min\left[\,1, \frac{p_c}{p_i}\,\right].$$

(2.12)

In the case of acceptance of a new point, we add it to the chain, otherwise it remains at the start point and a different candidate point is chosen. Once the jump is made and the new point is accepted, it is treated as the new start point and the steps are repeated until we have a large enough chain. MCMC methods are widely used for parameter inference, but it is inefficient in computing Bayesian evidence. Various nested sampling techniques/algorithms have been introduced for computing Bayesian evidence which include Skilling algorithm [12,13] like Multinest and CosmoNest. In this chapter, we use a Python interface for MultiNest [16–18], i.e., a PyMultiNest [19] code, and nested sampling [12], a generic Bayesian inference tool used to calculate Bayesian evidence. These techniques are used for constraining various cosmological models as done in recent research papers [20–23]. In the following, we present a detailed example of the application of Bayesian inference in Cosmology.

2.3 COSMOLOGICAL MODEL WITH INTERACTING DARK SECTOR

In cosmology, ΛCDM (cosmological constant + Cold Dark Matter) is the standard model. It is very successful as it finds a nice fit to observational data from different sources, yet it suffers from different problems [24]. In the ΛCDM model, dark energy is mimicked by the cosmological constant Λ with equation of state parameter $w_{de} = -1$. It does not interact with the CDM. Here we consider a cosmological model in which dark energy is different from the cosmological constant with the constant equation of state parameter w_{de} (not necessarily equal to -1), and it interacts with the dark matter via the interaction term Q, modeled by the following equations [25]:

$$\dot{\rho}_{dm} + 3\frac{\dot{a}}{a}\rho_{dm} = Q,$$

(2.13)

$$\dot{\rho}_{de} + 3\frac{\dot{a}}{a}(1 + w_{de})\rho_{de} = -Q.$$

(2.14)

Here ρ_{dm} and ρ_{de} are respectively the energy densities of dark matter and dark energy; a is the scale factor of the universe and an overdot stands for cosmic time t derivative. We will refer this model to as Interacting Dark Matter and Dark Energy (IDMDE) model. Obviously, it provides the standard ΛCDM scenario when $Q = 0$ and $w_{de} = -1$. In particular, we consider the model with the interaction term: $Q = H\delta\rho_{de}$, where $H = \frac{\dot{a}}{a}$ is the Hubble parameter. Solving the above equations for this interaction term, we get

$$\rho_{de} = \rho_{de0}a^{-[3(1 + w_{de}) + \delta]}, \tag{2.15}$$

$$\rho_{dm} = -\frac{\rho_{de0}\delta a^{-[3(1 + w_{de}) + \delta]}}{3w_{de} + \delta} + \left(\rho_{dm0} + \frac{\delta\rho_{de0}}{3w_{de} + \delta}\right)a^{-3}. \tag{2.16}$$

Here ρ_{dm0} represents the present-day energy density of dark matter, and ρ_{de0} represents the present-day energy density of dark energy. Therefore, we calculate overall energy density of the Universe as

$$\rho = \rho_m + \rho_{de} = \rho_{de0}a^{-[3(1 + w_{de} + \delta)]} - \frac{\rho_{de0}\delta a^{-[3(1 + w_{de}) + \delta]}}{3w_{de} + \delta} + \left(\rho_{dm0} + \frac{\delta\rho_{de0}}{3w_{de} + \delta}\right)a^{-3}. \tag{2.17}$$

Separating ρ_{dm0} and ρ_{de0} terms, we get

$$\rho = \rho_{dm0}a^{-3} + \rho_{de0}F(a)$$

where

$$F(a) = a^{-[3(1 + w_{de}) + \delta]} + \frac{a^{-[3(1 + w_{de}) + \delta]}}{3w_{de} + \delta} + \frac{\delta a^{-3}}{3w_{de} + \delta} \tag{2.18}$$

We represent the IDMDE model using Friedmann equation as

$$H^2 = \frac{8\pi G}{3}\left(\rho_{dm0}a^{-3} + \rho_{de0}F(a)\right) \tag{2.19}$$

In terms of density parameters $\Omega_{i0} = \frac{8\pi G\rho_{i0}}{3H_0^2}$, it reads as

$$H^2 = H_0^2\left(\Omega_{r0}a^{-4} + \Omega_{dm0}a^{-3} + \Omega_{de0}F(a)\right) \tag{2.20}$$

Note that when $a = 1$, $H = H_0$ and $\Omega_{dm0} + \Omega_{de0} = 1$.

2.4 DATA AND LIKELIHOODS

Following [26], we study the recent observational datasets, likelihoods and methodology used in our study for constraining the model.

2.4.1 $H(z)$

In this dataset, we consider 36 measurements of $H(z)$ as given in Table 2.2, the initial 31 $H(z)$ measurements are extracted via cosmic chronometric (CC) method, the other three measurements corresponding to $z = 0.38, 0.51, 0.61$ are derived from Baryon Acoustic Oscillations signal in galaxy distribution, finally the measurements $z = 2.34, 2.36$ are derived from Baryon Acoustic Oscillations signal in Lymann-α forest distribution. Let us consider 33 $H(z)$ measurements, i.e., combination of CC and Lyα. Chi-squared function represented by $\chi^2_{\text{CC + Ly}\alpha}$ is given by

$$\chi^2_{\text{CC + Ly}\alpha} = \sum_{i=1}^{33} \frac{[H^{\text{obs}}(z_i) - H^{\text{th}}(z_i)]^2}{\sigma^2_{H^{\text{obs}}(z_i)}}. \tag{2.21}$$

Here $H^{\text{obs}}(z_i)$ denotes the observed Hubble measurements, $\sigma^2_{H^{\text{obs}}(z_i)}$ is the standard deviation as shown in Table 2.2, and the theoretical Hubble measurements are obtained from the cosmological model, denoted by $H^{\text{th}}(z_i)$.

Corresponding to the measurements $z = 0.38, 0.51, 0.61$, we use the following covariance matrix:

$$C = \begin{bmatrix} 3.65 & 1.78 & 0.93 \\ 1.78 & 3.65 & 2.20 \\ 0.93 & 2.20 & 4.45 \end{bmatrix}. \tag{2.22}$$

Its chi-squared function is hence evaluated as

$$\chi^2_{Galaxy} = M^T C^{-1} M, \tag{2.23}$$

where

$$M = \begin{bmatrix} H^{\text{obs}}(0.38) - H^{\text{th}}(0.38) \\ H^{\text{obs}}(0.51) - H^{\text{th}}(0.51) \\ H^{\text{obs}}(0.61) - H^{\text{th}}(0.61) \end{bmatrix}. \tag{2.24}$$

Table 2.2 Hubble measurements

z_i	$H^{obs}(z_i)$ [km s^{-1} Mpc^{-1}]	$\sigma_{H^{obs}(z_i)}$
0.07	69	19.6
0.09	69	12
0.12	68.6	26.2
0.17	83	8
0.179	75	4
0.199	75	5
0.2	72.9	29.6
0.27	77	14
0.28	88.8	36.6
0.352	83	14
0.38	81.9	1.9
0.3802	83	13.5
0.4	95	17
0.4004	77	10.2
0.4247	87.1	11.2
0.4497	92.8	12.9
0.47	89	50
0.4783	80.9	9
0.48	97	62
0.51	90.8	1.9
0.593	104	13
0.61	97.8	2.1
0.68	92	8
0.781	105	12
0.875	125	17
0.88	90	40
0.9	117	23
1.037	154	20
1.3	168	17
1.363	160	33.6
1.43	177	18
1.53	140	14
1.75	202	40
1.965	186.5	50.4
2.34	223	7
2.36	227	8

Hence, the total chi-squared function χ_H^2 for $H(z)$ measurements is

$$\chi_H^2 = \chi_{CC+Ly\alpha}^2 + \chi_{Galaxy}^2. \tag{2.25}$$

2.4.2 Baryon acoustic oscillations (BAO)

These measurements are widely used to study the angular diameter distance which is evaluated in terms of Hubble parameter and its redshift function. The measurements are represented in terms of redshift separation and angular scale. $d(z)$ is the dimensionless ratio evaluated as

$$d(z) = \frac{r_s(z_d)}{D_V(z)}. \tag{2.26}$$

The comoving size $r_s(z_d)$ at the drag redshift $z_d = 1059.6$ of the sound horizon is calculated as

$$r_s(z_d) = \int_{z_d}^{\infty} \frac{c_s dz}{H(z)}. \tag{2.27}$$

Here, $c_s = \frac{c}{\sqrt{3(1+\mathscr{R})}}$ denotes the baryon-photon fluid's sound speed, and $\mathscr{R} = \frac{3\Omega_{b0}}{4\Omega_{r0}(1+z)}$ with $\Omega_{b0} = 0.022h^{-2}$ and $\Omega_{r0} = \Omega_{\gamma 0}\left(1 + \frac{7}{8}\left(\frac{4}{11}\right)^{\frac{4}{3}}N_{eff}\right)$, where $\Omega_{\gamma 0} = 2.469 \times 10^{-5}h^{-2}$ and $N_{eff} = 3.046$.

Furthermore, the relation between transverse distance scale and the line of sight is given by the volume averaged distance $D_V(z)$:

$$D_V(z) = \left[(1+z)^2 D_A(z)^2 \frac{cz}{H(z)}\right]^{1/3}. \tag{2.28}$$

Here, the angular diameter distance is given by $D_A(z) = \frac{c}{1+z}\int_0^z \frac{dz}{H(z)}$, where c is the speed of light.

Let us consider the first five measurements as represented in Table 2.3 from different surveys, the corresponding chi-squared function is

$$\chi_{NW}^2 = \sum_{i=1}^{5}\left[\frac{d^{obs}(z_i) - d^{th}(z_i)}{\sigma_{d(z_i)}}\right]^2. \tag{2.29}$$

Here $d^{obs}(z_i)$ denotes the dimensionless ratios observed value with the deviation $\sigma_{d(z_i)}$ as shown in Table 2.3, and $d^{th}(z_i)$ represents the theoretical measurement calculated from the respective model.

Table 2.3 BAO data

Survey	z_i	$d(z_i)$	$\sigma_{d(z_i)}$
6dFGS	0.106	0.3360	0.0150
MGS	0.15	0.2239	0.0084
BOSS LOWZ	0.32	0.1181	0.0024
SDSS(R)	0.35	0.1126	0.0022
BOSS CMASS	0.57	0.0726	0.0007
WiggleZ	0.44	0.073	0.0012
WiggleZ	0.6	0.0726	0.0004
WiggleZ	0.73	0.0592	0.0004

Finally, the three measurements from the WiggleZ survey are investigated with the help of the inverse covariance matrix

$$C^{-1} = \begin{bmatrix} 1040.3 & -807.5 & 336.8 \\ -807.5 & 3720.3 & -1551.9 \\ 336.8 & -1551.9 & 2914.9 \end{bmatrix}. \tag{2.30}$$

Chi-squared function(χ_W^2), for the WiggleZ data is given by

$$\chi_W^2 = D^T C^{-1} D, \tag{2.31}$$

where

$$D = \begin{bmatrix} d^{obs}(0.44) - d^{th}(0.44) \\ d^{obs}(0.6) \;\; - d^{th}(0.6) \\ d^{obs}(0.73) - d^{th}(0.73) \end{bmatrix}. \tag{2.32}$$

So, total BAO contribution chi-squared function or χ_{BAO}^2, reads as

$$\chi_{BAO}^2 = \chi_{NW}^2 + \chi_W^2. \tag{2.33}$$

2.4.3 Cosmic microwave background

In our analysis, we use the compressed Cosmic Microwave Background likelihood of Planck 2015 data. The angular scale l_a at the last scattering surface of the sound horizon is formulated as

$$l_a = \pi \frac{r(z_*)}{r_s(z_*)}. \tag{2.34}$$

At the last scattering surface, we evaluate the comoving distance $r(z_*)$ as

$$r(z_*) = \int_0^{z_*} \frac{cdz}{H(z)}, \tag{2.35}$$

where the comoving sound horizon $r_s(z_*)$ is computed at $z_* = 1089.9$ (last scattering's redshift).

The chi-squared function χ^2_{CMB} of CMB is represented as

$$\chi^2_{CMB} = \frac{(l_a^{obs} - l_a^{th})^2}{\sigma_{l_a}^2}, \tag{2.36}$$

where $l_a^{obs} = 301.63$ denotes the sound horizon's angular scale's observed value, having an uncertainty $\sigma_{l_a} = 0.15$, and l_a^{th} denotes the theoretical measurement determined from the considered model.

2.4.4 Methodology

We use the combined combination of $H(z)$, BAO, and CMB data for constraining the IDMDE and ΛCDM model parameters. The analysis is done through PyMultiNest code, and nested sampling technique as mentioned before. The joint multivariate Gaussian likelihood $\mathscr{L}_{Joint}$ used is as follows:

$$\mathscr{L}_{Joint} \propto \exp\left(\frac{-\chi^2_{Joint}}{2}\right) \tag{2.37}$$

Here,

$$\chi^2_{Joint} = \chi^2_H + \chi^2_{BAO} + \chi^2_{CMB}. \tag{2.38}$$

is the chi-squared function representing the combination of datasets mentioned above.

For all the model parameters in our analysis, we use the uniform prior distribution, viz., $55 \le H_0 \le 85$, $0.1 \le \Omega_{dm0} \le 0.5$, $0 \le \Omega_{\sigma 0} \le 0.1$, $-0.1 \le \delta \le 0.1$, and $-2 \le w_{de} \le 0$, respectively.

2.5 RESULTS AND DISCUSSION

Table 2.4 summarizes the results of our analysis wherein 68%, 95%, and 99% confidence limits (CL) constraints are displayed on the interaction model parameters from H(z)+BAO+CMB data. The Hubble constant H_0 should be read in units of km s^{-1} Mpc^{-1}. Meanwhile, Figure 2.1 displays one- and two-dimensional (68%, 95%, 99% CLs) marginalized probability distributions of the model parameters. First, we focus our attention on the main parameter of interest, which is the interaction parameter δ. We see that δ in non-zero at 95% CL in the range [0.0004,0.0188], which in turn implies that the observational data offers strong evidence for the interaction between dark energy and dark matter within the framework of the model under consideration.

From Figure 2.1, we see that δ shows a negative correlation with H_0. That means larger values of δ are related to smaller values of H_0. That is why we observe a smaller mean value of H_0 in the IDMDE model when compared to the ΛCDM model, since the mean value of δ is larger in the IDMDE model. In contrast, we observe a positive correlation of δ with the other two parameters w_{de} and Ω_{dm0}. That means larger values of δ are related to larger values of w_{de} and Ω_{dm0}. The values of the EoS parameter w_{de} of dark energy are greater than –1 in 68% confidence interval showing the quintessence behavior of dark energy at this CL.

From Table 2.4 and Figure 2.1, we notice tight constraints on the free parameters of the ΛCDM model compared to the IDMDE model. This is because the IDMDE model carries three extra parameters, and is therefore penalized in the statistical fitting with the data. It may also be observed from the Bayesian evidence of the two models that there is strong evidence against the IDMDE model in comparison to the Λ model as per Jeffery's criterion given in Table 2.1.

Thus, in the above example of a cosmological model, we have studied an extension of the standard model of cosmology by the interaction between

Table 2.4 The 68%, 95% and 99% CL constraints on the model parameters from H(z) +BAO+CMB data. The parameter H_0 is measured in units of km s^{-1} Mpc^{-1}

Model	IDMDE	ΛCDM
H_0	$66.4^{+2.2+4.3+5.9}_{-2.2-4.2-5.3}$	$68.51^{+0.45+0.87+1.1}_{-0.45-0.87-1.1}$
Ω_{dm0}	$0.256^{+0.010+0.019+0.026}_{-0.010-0.020-0.025}$	$0.2539^{+0.0054+0.011+0.014}_{-0.0054-0.010-0.013}$
w_{de}	$-0.905^{+0.089+0.17+0.22}_{-0.089-0.18-0.24}$	-1
δ	$0.0093^{+0.0040+0.0095+0.015}_{-0.0050-0.0089-0.010}$	0
BE	-31.21	-27.40

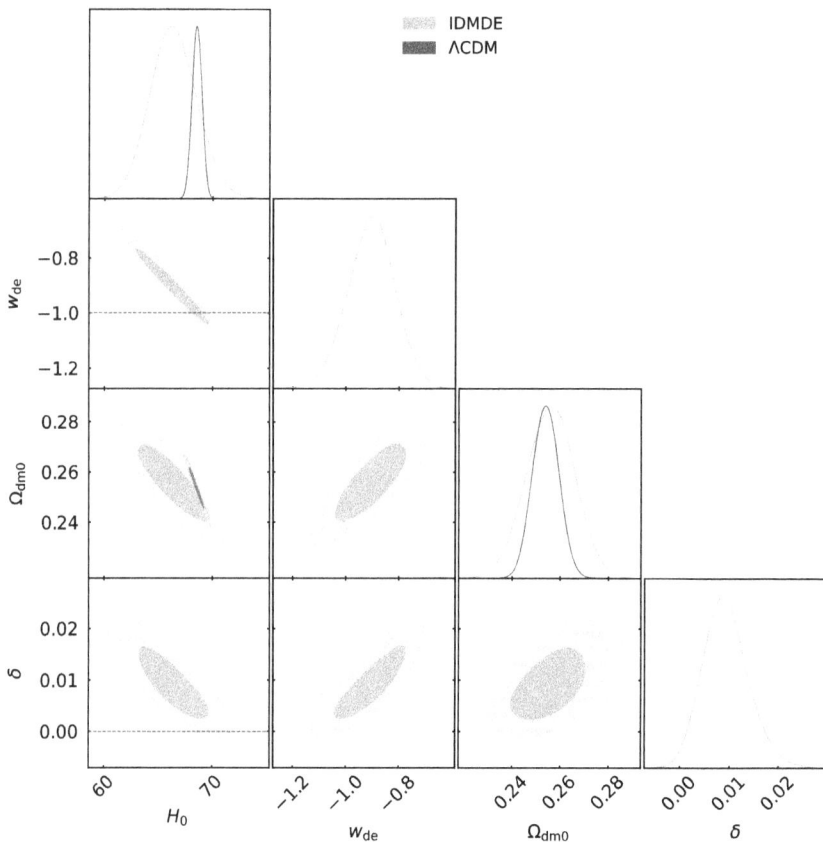

Figure 2.1 One- and two-dimensional (68%, 95%, 99% CLs) marginalized distributions of
the model parameters.

dark matter and dark energy. We have constrained the model with
$H(z)$+BAO+CMB data and found that the interaction parameter δ in
non-zero at 95% CL in the range [0.0004,0.0188], which in turn implies
that the observational data offers strong evidence for the interaction
between dark energy and dark matter within the framework of the model
under consideration. Also, a positive value of δ in this model implies that
transfer of energy takes place from dark energy to dark matter. The model
investigated here is interesting, and deserves further investigation with more
observational data from different sources.

2.6 CONCLUSION

We have presented the basics of Bayesian statistics, along with the MCMC
method, and demonstrated its usage via studying a cosmological model.

REFERENCES

1. R. Trotta, "Bayesian methods in cosmology", *arXiv:1701.01467*, 2017.
2. L. E. Padilla *et al.*, "Cosmological parameter inference with Bayesian statistics", *arXiv:1903.11127*, 2019.
3. B. Santos, N. C. Devi and J. S. Alcaniz, "Bayesian comparison of nonstandard cosmologies using type Ia supernovae and BAO data", Physical Review D, vol. 95, no. 12, p. 123514, 2017.
4. M. A. Santos, M. Benetti, J. Alcaniz, F. A. Brito and R. Silva, "CMB constraints on β-exponential inflationary models", Journal of Cosmology and Astroparticle Physics, vol. 03, p. 023, 2018.
5. S. Santos da Costa, M. Benetti and J. Alcaniz, "A Bayesian analysis of inflationary primordial spectrum models using Planck data", Journal of Cosmology and Astroparticle Physics, vol. 03, p. 004, 2018.
6. U. Andrade, C. A. P. Bengaly, J. S. Alcaniz and B. Santos, "Isotropy of low redshift type Ia supernovae: A Bayesian analysis", Physical Review D, vol. 97, no. 8, p. 083518, 2018.
7. A. Cid, B. Santos, C. Pigozzo, T. Ferreira and J. Alcaniz, "Bayesian comparison of interacting scenarios", Journal of Cosmology and Astroparticle Physics, vol. 03, p. 030, 2019.
8. R. Trotta, "Bayes in the sky: Bayesian inference and model selection in cosmology", Contemporary Physics, vol. 49, no. 2, p. 71, 2008.
9. H. Jeffreys, Theory of probability, Oxford Classics Series, 1998.
10. N. Metropolis, A. Rosenbluth, M. Rosenbluth, A. Teller and E. Teller, "Equations of state calculations by fast computing machines", The journal of chemical physics, vol. 21, no. 6, p. 1087, 1953.
11. W. K. Hastings, "Monte Carlo sampling methods using Markov chains and their applications", p. 97, 1970.
12. J. Skilling, "Nested sampling", AIP Conference Proceedings, vol. 735, p. 395, 2004.
13. J. Skilling, "Nested sampling for general Bayesian computation", Bayesian Analysis, vol. 1, no. 4, p. 833, 2006.
14. S. Brooks, A. Gelman, G. Jones and X.-L. Meng, Handbook of Markov chain Monte Carlo. CRC Press, 2011.
15. C. P. Robert and G. Casella, "The Metropolis-Hastings algorithm", Monte Carlo Statistical Methods, p. 231, Springer, 1999.
16. F. Feroz and M. P. Hobson, "Multimodal nested sampling: An efficient and robust alternative to Markov Chain Monte Carlo methods for astronomical data analyses", Monthly Notices of the Royal Astronomical Society, vol. 384, no. 2, p. 449, 2008.
17. F. Feroz, M. P. Hobson and M. Bridges, "MultiNest: An efficient and robust Bayesian inference tool for cosmology and particle physics", Monthly Notices of the Royal Astronomical Society, vol. 398, no. 4, p. 1601, 2009.
18. F. Feroz, M. P. Hobson, E. Cameron and A. N. Pettitt, "Importance nested sampling and the MultiNest algorithm", *arXiv:1306.2144*, 2013.
19. J. Buchner *et al.*, "X-ray spectral modelling of the AGN obscuring region in the CDFS: Bayesian model selection and catalogue", Astronomy & Astrophysics, vol. 564, p. A125, 2014.

20. Ö. Akarsu, S. Kumar, S. Sharma and L. Tedesco, "Constraints on Bianchi type-I spacetime extension of the standard ΛCDM model", Physical Review D, vol. 100, p. 023532, 2019.
21. S. Sharma, "Probing phenomenological emergent dark energy model in a Bianchi type-I spacetime with the recent observational data", Physics of the Dark Universe, p. 100717, 2020.
22. Ö. Akarsu, N. Katrc, S. Kumar, R. C. Nunes, B. Öztürk and S. Sharma, "Rastall gravity extension of the standard Λ CDM model: theoretical features and observational constraints." The European Physical Journal C, vol. 80, p. 1, 2020.
23. Ö. Akarsu, E. Di Valentino, S. Kumar, M. Ozyigit and S. Sharma, "Testing spatial curvature and anisotropic expansion on top of the Λ CDM model." arXiv:2112.07807, 2021.
24. S. Dodelson, Modern Cosmology. Elsevier, 2003.
25. W. Yang, S. Pan, R. C. Nunes and D. F. Mota, "Dark calling dark: Interaction in the dark sector in presence of neutrino properties after Planck CMB final release", Journal of Cosmology and Astroparticle Physics, vol. 04, p. 008, 2020.
26. O. Akarsu, S. Kumar, S. Sharma and L. Tedesco, "Constraints on a Bianchi type I spacetime extension of the standard ΛCDM model", Physical Review D, vol. 100, p. 023532, 2019.

Chapter 3

An improved sufficient bootstrapping

Hamed Olalekan, Stephen A. Sedory, and Sarjinder Singh

3.1 INTRODUCTION

According to the Oxford Dictionary, "to bootstrap" means to get something into or out of a situation using existing resources. In its statistical usage, "Bootstrapping" is a scheme that uses sampling in a random method with replacement that belongs to the larger class of sampling techniques and called resampling methods. The conventional bootstrapping scheme provides measures of accuracy to estimates. This bootstrapping technique makes possible in approximation of the sampling distribution of almost any estimator of a population parameter that uses sampling methods that are deemed random. A recent study related to the conventional bootstrapping method, is the paper by Singh and Sedory [1] that introduces sufficient bootstrapping. The authors proposed the idea of using distinct (or unique) individual units when using bootstrapping to estimate a parameter. Their results suggested that using the newly developed sufficient bootstrap sampling method could reduce computational time and that the method also results in a gain in relative efficiency when compared to conventional bootstrap for certain cases.

Growth in technology saw a significant increase in the early 90s, consequently, computer dependent solutions became increasingly popular in the literature of statistics. The Jackknife-after-bootstrap (JaB) sampling technique which was developed by Effron [2], is one of these computer-dependent solutions. It has particularly been used by Beyaztas and Alin [3]. They used it to ascertain the cut-off values for various linear regression diagnostic models and were able to obtain more accurate solutions than the non-conventional methods, especially under non-normal errors as well as with small samples. Beyaztas and Alin [3] also compared the sufficient bootstrap and Jackknife bootstrap methods and showed that the concept proposed in sufficient bootstrapping can easily be applied to the JaB method. They also compared it to the conventional bootstrapping methods, recognizing that the downside of the JaB method is that it requires a lot of computational time. They further proved that by incorporating sufficient

DOI: 10.1201/9781003356653-3

bootstrapping sampling method into JaB, the same accuracy can be maintained using less computation. Furthermore, they also showed that the positive solutions derived from the sufficient bootstrap sampling method and conventional JaB sampling method have proven effective in detecting influential samples in linear regression solutions. Comparison of bootstrapping methods was done using two real-world samples and an extensive simulation design study. The investigated design combines different sample sizes with diverse modeling scenarios. The results showed that the proposed sampling method is a better alternative than the conventional JaB method with a significantly lesser standard error and computation time. For further studies, refer to sufficient Jackknife-after-bootstrap method for detection of influential observations in linear regression models by Beyaztas and Alin [4].

Alin et al. [5] combined sufficient bootstrapping with the idea of m/n bootstrapping sampling so as to regain stable estimation of the distribution of the samples and to further reduce the burden of computation using the bootstrap. Beyaztas and Alin [3] gathered all necessary and sufficient conditions for asymptotic normality of the method, and also came up with new values for the resample size m. They compared the proposed method to the initial traditional bootstrap, to sufficient bootstrap sampling method and to the m/n bootstrap using simulation samples. The results produced were a positive gain in terms of efficiency and of the amount of computational time. For further studies please refer to "Sufficient m-out-of-n (m/n) bootstrap" by Beyaztas, Alin and Bandyopadhyay [6].

Dérian et al. [7] used sufficient bootstrapping of the values of transcriptome signatures from immunized mouse dendritic cells to predict late vaccine-induced t-cell responses. Beyaztas [8] proposed the sufficient time series bootstrap sampling method and his results show that it achieves a better output than the traditional non-overlapping block-bootstrap; the results also show less computing time and much lower standard errors than the traditional non-overlapping block-bootstrap.

Beyaztas and Beste [9] in their study used sufficient bootstrap sampling method and ordered non-overlapping block bootstrap methods into an algorithm system that followed the Jackknife-after-bootstrap (JaB) and they used it to determine the standard error of the statistics sampled where the sample observations form a sequence that is stationary. They further extended the JaB algorithm so as to get estimated intervals for future returns and for volatilities of GARCH processes by using "sufficient bootstrap" method by Singh and Sedory [1]. This method only uses distinct units in a bootstrap sample in order to reduce the computational burden for larger sample sizes, sampling distributions is also consistent based on the obtained results from the GARCH process as shown in Beyaztas et al. [10].

A study of the assessment of individual bioequivalence using sufficient bootstrap procedure by Beyaztas [11] proposed a sufficient bootstrap method, which uses only distinct samples and observations to assess the individual bioequivalence under 2×4 randomized crossover design. The proposed method is shown by using an extensive Monte Carlo simulation study as well as a real sample data set, and the outputs are compared with the output obtained from the conventional bootstrap sampling technique. The result records show that the proposed method is a better alternative and can pass as a better result than the conventional bootstrap confidence limits percentile.

Mostafa and Ahmed [12] compared numerous block bootstrap sampling methods in terms of bias of estimation of parameters as well as in terms of the mean squared errors (MSE) of the bootstrapping technique sample estimators. The comparison is based on four real-world samples and on a comprehensive simulation study involving numerous sample sizes and parameters. The results from the simulation showed that the ordered and sufficient bootstrap sampling technique and the ordered non-overlapping block bootstrap methods proposed by Beyaztas [8] provide better results in terms of parameter estimation and its MSE compared to conventional methods.

The chapter proceeds as follows. Section 3.2 is devoted to discussing conventional bootstrapping; Section 3.3 discusses sufficient bootstrapping; Section 3.4 introduces improved sufficient bootstrapping; Section 3.5 focuses on the theoretical developments of the three types of bootstrapping; Section 3.6 deals with analytical comparisons; and Section 3.7 demonstrates the performance of the proposed improved sufficient bootstrapping method through an extensive simulation study. A conclusion ends the chapter.

3.2 CONVENTIONAL BOOTSTRAPPING

Consider a simple random sampling with replacement (SRSWR) sample of $n = 3$ units taken from a population of $N = 3$ units, say Y_1, Y_2 and Y_3 taking values of 12, 15 and 21, respectively. Obviously, the total number of SRSWR samples will be N^n, where N is the population size and n is the sample size that is $3^3 = 27$ samples. Then the conventional bootstrap can be considered as a unique case of SRSWR where the sample size n becomes equal to the population size N. We can easily compute the population mean, population standard deviation, and population coefficient of variation, respectively, as

$$\mu_y = 16 \tag{3.1}$$

$$\sigma_y = 3.74 \tag{3.2}$$

and

$$C_y = \frac{\sigma_y}{\mu_y} = 0.2339 \tag{3.3}$$

In the next three sections, we compare the results of the three types of bootstrapping using the above sample of 3 units. We illustrate how the methods differ by considering one particular set of possible samples namely: (Y_1, Y_1, Y_2); (Y_1, Y_2, Y_1) and (Y_2, Y_1, Y_1). Conventional bootstrapping treats these as expected: $y_1 = Y_1$, $y_2 = Y_1$, $y_3 = Y_2$ for the first, $y_1 = Y_1$, $y_2 = Y_2$, $y_3 = Y_1$ for the second, and $y_1 = Y_2$, $y_2 = Y_1$, $y_3 = Y_1$ for the third. We note that for the outcome (Y_2, Y_2, Y_2) there is clearly only one possibility. Sufficient Bootstrapping (SB) would treat these three outcomes as samples of sizes 2 namely $y_1 = Y_1$, $y_2 = Y_2$ for the first and second, and $y_1 = Y_2$, $y_2 = Y_1$ for the third. Improved Sufficient Bootstrapping (ISB) retains the sample size of 3 but treats the outcomes as $y_1 = Y_1$, $y_2 = \bar{Y}$, $y_3 = Y_2$ for the first; $y_1 = Y_1$, $y_2 = Y_2$, $y_3 = \bar{Y}$ for the second; and $y_1 = Y_2$, $y_2 = Y_1$, $y_3 = \bar{Y}$ for the third. Again for the outcome (Y_2, Y_2, Y_2), SB treats it as a sample of size 1, with $y_1 = Y_2$; the ISB could treat it as $y_1 = Y_2$, $y_2 = \bar{Y}$, $y_3 = \bar{Y}$. Conventional Bootstrapping (CB) would give the results as given in the following Table 3.1.

The expected value of the CB mean is given by

$$E(\bar{y}_{CB}) = \sum_{CB} P_{CB}^{(1)} \bar{y}_{CB} = 16 = \mu_y \tag{3.4}$$

Thus, we see clearly that the sample mean is an unbiased estimator of the population mean in the CB sample (Tables 3.2 and 3.3).

The variance of the CB mean is given by

$$V(\bar{y}_{CB}) = \sum_{CB} P_{CB}^{(1)} [\bar{y}_{CB} - E(\bar{y}_{CB})]^2 = 4.67 \tag{3.5}$$

Table 3.1 Frequency distribution table for mean of CB

Sample mean $\bar{y}_{CB}$	12	13	14	15	16	17	18	19	21	Sum
Frequency	1	3	3	4	6	3	3	3	1	27
$P_{CB}^{(1)}$	1/27	3/27	3/27	4/27	6/27	3/27	3/27	3/27	1/27	1

Table 3.2 Frequency distribution for variance of CB

Sample variance (s_{CB}^2)	0	3	12	21	27	Sum
Frequency (f)	3	6	6	6	6	27
$P_{CB}^{(2)}$	3/27	6/27	6/27	6/27	6/27	1

Table 3.3 Frequency distribution of standard deviation for CB

Sample standard deviation sd_{CB}	0	1.73	3.46	4.58	5.20	Sum
Frequency (f)	3	6	6	6	6	27
$P_{CB}^{(3)}$	3/27	6/27	6/27	6/27	6/27	1

The expected value of the variance of CB is given by

$$E\left(s_{CB}^2\right) = \sum_{CB} P_{CB}^{(2)} s_{CB}^2 = 14. \tag{3.6}$$

The value of CB variance is given by

$$V\left(s_{CB}^2\right) = \sum_{CB} P_{CB}^{(2)} \left[s_{CB}^2 - E\left(s_{CB}^2\right)\right]^2 = 98 \tag{3.7}$$

The expected value of CB sample standard deviation is given by

$$E(sd_{CB}) = \sum_{CB} P_{CB}^{(3)} S_{CB} = 3.33 \tag{3.8}$$

The mean square error (MSE) of CB sample standard deviation is given by

$$MSE(sd_{CB}) = \sum_{CB} P_{CB}^{(3)} \left[sd_{CB} - \sigma_y\right]^2 = 3.09 \tag{3.9}$$

From Table 3.4 the expected value of CB sample coefficient of variation is given by:

$$E(\hat{C}_{CB}) = \sum_{CB} P_{CB}^{(4)} \hat{C}_{CB} = 0.2057 \tag{3.10}$$

The MSE of CB sample coefficient of variation is given by

$$MSE(\hat{C}_{CB}) = \sum_{CB} P_{CB}^{(4)} \left[\hat{C}_{CB} - C_y\right]^2 = 0.0113 \tag{3.11}$$

Table 3.4 Frequency distribution for coefficient of variation of CB

Sample CV, $\hat{C}_{CB}$	0	0.1237	0.1332	0.1823	0.2038	0.2864	0.2887	0.3464	Sum
Frequency (f)	3	3	3	3	3	6	3	3	27
$P_{CB}^{(4)}$	3/27	3/27	3/27	3/27	3/27	6/27	3/27	3/27	1

The relative bias in CB sample standard deviation is given by

$$\mathrm{RB}(sd_{CB}) = \frac{\mathrm{E}(sd_{CB}) - \sigma_y}{\sigma_y} \times 100\% = -10.9626\% \qquad (3.12)$$

Thus, CB underestimates the standard deviation.

Also, in this case, the relative bias in CB sample coefficient of variation is given by

$$\mathrm{RB}(\hat{C}_{CB}) = \frac{\mathrm{E}(\hat{C}_{CB}) - C_y}{C_y} \times 100\% = -12.056\% \qquad (3.13)$$

In this example, CB also underestimates the coefficient of variation.

In the next section, we explain the concept of Sufficient Bootstrapping (SB) with a numerical illustration.

3.3 SUFFICIENT BOOTSTRAPPING

Next, considering the same 27 possible outcomes as before, but regarding these as samples under the SB scheme in the way introduced in Section 3.2, we computed the corresponding tables found below:

From Table 3.5 the expected value of SB mean is given by

$$\mathrm{E}(\bar{y}_{SB}) = \sum_{SB} P_{SB}^{(1)} \bar{y}_{SB} = 16 = \mu_y \qquad (3.14)$$

Thus, in this case SB sample mean is also an unbiased estimator of the population mean.

The variance of SB mean is given by

$$\mathrm{V}(\bar{y}_{SB}) = \sum_{SB} P_{SB}^{(1)} \left[\bar{y}_{SB} - \mathrm{E}(\bar{y}_{SB}) \right]^2 = 3.889 \qquad (3.15)$$

Table 3.5 Frequency distribution for mean of SB

Sample mean $\bar{y}_{SB}$	12	13.5	15	16	16.5	18	21	Sum
Frequency (f)	1	6	1	6	6	6	1	27
$P_{SB}^{(1)}$	1/27	6/27	1/27	6/27	6/27	6/27	1/27	1

Table 3.6 Frequency distribution of variance for SB

Sample variance (s_{SB}^2)	0	4.5	18	21	40.5	Sum
Frequency (f)	3	6	6	6	6	27
$P_{SB}^{(2)}$	3/27	6/27	6/27	6/27	6/27	1

From Table 3.6 the expected value of SB variance is given by

$$E(s_{SB}^2) = \sum_{SB} P_{SB}^{(2)} s_{SB}^2 = 18.7 \tag{3.16}$$

The mean square of SB variance is given by

$$\text{MSE}\left(s_{SB}^2\right) = \sum_{SB} P_{SB}^{(2)} \left[s_{SB}^2 - \sigma_y^2\right]^2 = 212.3 \tag{3.17}$$

From Table 3.7 the expected value of SB sample standard deviation is given by

$$E(sd_{SB}) = \sum_{SB} P_{SB}^{(3)} s_{SB} = 3.85 \tag{3.18}$$

The mean square error of SB standard deviation is given by

$$\text{MSE}(sd_{SB}) = \sum_{SB} P_{SB}^{(3)} [s_{SB} - \sigma_y]^2 = 3.88 \tag{3.19}$$

From Table 3.8, the expected value of SB sample coefficient of variation is given by

$$E(\hat{C}_{SB}) = \sum_{SB} P_{SB}^{(4)} \hat{C}_{SB} = 0.240 \tag{3.20}$$

Table 3.7 Frequency distribution of standard deviation for SB

Sample standard deviation s_{SB}	0	2.12	4.24	4.58	6.36	Sum
Frequency (f)	3	6	6	6	6	27
$P_{SB}^{(3)}$	3/27	6/27	6/27	6/27	6/27	1

Table 3.8 Frequency distribution of coefficient of variation for SB

Sample CV, $\hat{C}_{SB}$	0	0.1571	0.2357	0.2864	0.3857	Sum
Frequency (f)	3	6	6	6	6	27
$P_{SB}^{(4)}$	3/27	6/27	6/27	6/27	6/27	1

The variance of SB coefficient of variation is given by

$$V(\hat{C}_{SB}) = \sum_{SB} P_{SB}^{(4)} \left[\hat{C}_{SB} - C_y \right]^2 = 0.013 \tag{3.21}$$

Thus, the relative bias in SB sample standard deviation is given by

$$RB(\hat{s}_{SB}) = \frac{E(S_{SB}) - \sigma_y}{\sigma_y} \times 100\% = 2.82\% \tag{3.22}$$

Also, the relative bias in SB sample coefficient of variation is given by

$$RB(\hat{C}_{SB}) = \frac{E(\hat{C}_{SB}) - C_y}{C_y} \times 100\% = -1.19\% \tag{3.23}$$

Interestingly for this particular example, we see by using Equation 3.9 and Equation 3.10 that the relative bias in the SB estimator of standard deviation is much less than that based on CB. Also note that SB over-estimates the standard deviation while CB underestimates the standard deviation. In addition, the estimator of the coefficient of variation based on SB has less relative bias than that based on CB method. Tables 3.1 and 3.5 show that the distributions of sample mean based on conventional and sufficient bootstrapping differ. From Tables 3.3 and 3.7 the distributions of sample standard deviations remain similar. Thus, SB is unbiased while estimating the population mean, overestimates the standard deviation, and underestimates the coefficient of variation in this particular example.

In Table 3.9, we summarize the numerical results for SB and CB.

Now we can easily calculate the relative efficiency of SB with respect to CB as follows.

The percent relative efficiency of SB over CB while considering the problem of estimating mean is given by

$$RE(\bar{y}_{SB}) = \frac{V(\bar{y}_{CB})}{V(\bar{y}_{SB})} \times 100\% = 120\% \tag{3.24}$$

Table 3.9 Comparing the variance (or MSE) of SB and CB

Parameters	CB	SB
Mean	4.67	3.889
Variance	98	212.3
Standard deviation	3.09	3.880
Coefficient of variation	0.0113	0.0132

The percent relative efficiency of SB over CB while considering the problem of estimating variance is given by

$$\text{RE}(s_{SB}^2) = \frac{V\left(s_{CB}^2\right)}{\text{MSE}\left(s_{SB}^2\right)} \times 100\% = 46.22\% \tag{3.25}$$

The percent relative efficiency of SB over CB while considering the problem of estimating standard deviation is given by

$$\text{RE}(sd_{SB})) = \frac{\text{MSE}\left(sd_{CB}\right)}{\text{MSE}\left(sd_{SB}\right)} \times 100\% = 79.64\% \tag{3.26}$$

The percent relative efficiency of SB over CB while considering the problem of estimating coefficient of variation is given by

$$\text{RE}(\hat{C}_{SB}) = \frac{\text{MASE}\left(\hat{C}_{CB}\right)}{\text{MSE}\left(\hat{C}_{SB}\right)} \times 100\% = 85.06\% \tag{3.27}$$

Clearly, we see and conclude that SB is more efficient while estimating mean than CB, but could perform less efficiently in certain cases as mentioned in Singh and Sedory [1]. Thus, in the next section, we suggest an improved sufficient bootstrapping (ISB) sampling scheme.

3.4 IMPROVED SUFFICIENT BOOTSTRAPPING

Finally in this section, we treat the 27 samples under the proposed ISB scheme. Table 3.10 lists each of the 27 samples explicitly as to how it is regarded and gives the corresponding sample statistics.

From Table 3.11, the expected value of ISB mean is given by

$$\text{E}(\bar{y}_{ISB}) = \sum_{ISB} P_{ISB}^{(1)} \bar{y}_{ISB} = 16 = \mu_y \tag{3.28}$$

Thus, in this case ISB sample mean is also an unbiased estimator of the population mean.

The variance of ISB mean is given by

$$\text{V}(\bar{y}_{ISB}) = \sum_{ISB} P_{ISB}^{(1)} \left[\bar{y}_{ISB} - \text{E}(\bar{y}_{ISB})\right]^2 = 1.210 \tag{3.29}$$

Table 3.10 Proposed method of ISB

Y1	Y2	Y3	Mean_ISB $\bar{y}_{ISB}$	VAR_ISB s^2_{ISB}	Distinct Units(u)	VAR_ADJ $s^2_{ISB(ADJ)}$	SD_ADJ $s_{ISB(ADJ)}$	CV_ADJ $\hat{C}_{ISB(ADJ)}$
12	16	16	14.667	5.333	1	7.111	2.667	0.182
12	16	15	14.333	4.333	2	2.944	1.716	0.120
12	16	21	16.333	20.333	2	20.278	4.503	0.276
12	15	16	14.333	4.333	2	2.944	1.716	0.120
12	15	16	14.333	4.333	2	2.944	1.716	0.120
12	15	21	16.000	21.000	3	14.000	3.742	0.234
12	21	16	16.333	20.333	2	20.278	4.503	0.276
12	21	15	16.000	21.000	3	14.000	3.742	0.234
12	21	16	16.333	20.333	2	20.278	4.503	0.276
15	12	16	14.333	4.333	2	2.944	1.716	0.120
15	12	16	14.333	4.333	2	2.944	1.716	0.120
15	12	21	16.000	21.000	3	14.000	3.742	0.234
15	16	12	14.333	4.333	2	2.944	1.716	0.120
15	16	16	15.667	0.333	1	0.444	0.667	0.043
15	16	21	17.333	10.333	2	9.444	3.073	0.177
15	21	12	16.000	21.000	3	14.000	3.742	0.234
15	21	16	17.333	10.333	2	9.444	3.073	0.177
15	21	16	17.333	10.333	2	9.444	3.073	0.177
21	12	16	16.333	20.333	2	20.278	4.503	0.276

21	12	15	16.000	21.000	3	14.000	3.742	0.234
21	12	16	16.333	20.333	2	20.278	4.503	0.276
21	15	12	16.000	21.000	3	14.000	3.742	0.234
21	15	16	17.333	10.333	2	9.444	3.073	0.177
21	15	16	17.333	10.333	2	9.444	3.073	0.177
21	16	12	16.333	20.333	2	20.278	4.503	0.276
21	16	15	17.333	10.333	2	9.444	3.073	0.177
21	16	16	17.667	8.333	1	11.111	3.333	0.189

where the abbreviations "ISB" and "ADJ" stand for ISB and adjustment in the ISB.

Table 3.11 Frequency distribution of mean for ISB

Mean $\bar{y}_{ISB}$	14.34	14.67	15.67	16	16.33	17.33	17.66	Sum
Frequency (f)	6	1	1	6	6	6	1	27
Probability $P_{ISB}^{(1)}$	6/27	1/27	1/27	6/27	6/27	6/27	1/27	1

Note that in the case of ISB, we suggest an adjusted estimator of variance given by

$$s_{ISB\,(ADJ)}^2 = \frac{(n-1)\,S_{ISB}^2 - \frac{(n-u)\,u^2\,(\bar{y}_n - \bar{y}_u)^2}{n^2}}{u} \tag{3.30}$$

where u is the number of distinct units in the SB. The proof of the adjustment in (3.30) is given in Theorem 5.3 in Section 3.5.

From Table 3.12, the expected value of the adjusted estimator of variance for ISB is given by

$$E\left(s_{ISB\,(ADJ)}^2\right) = \Sigma_{ISB}\ P_{ISB}^{(2)} s_{ISB\,(ADJ)}^2 = 11.06 \tag{3.31}$$

The MSE of the adjusted estimator of variance is given by

$$\text{MSE}\left(s_{ISB\,(ADJ)}^2\right) = \Sigma_{ISB}\ P_{ISB}^{(2)}\left[s_{ISB\,(ADJ)}^2 - \sigma_y^2\right]^2 = 49.40 \tag{3.32}$$

From Table 3.13 the expected value of the adjusted standard deviation for ISB is given by

$$E\left(sd_{ISB\,(ADJ)}\right) = \Sigma_{ISB}\ P_{ISB}^{(3)} sd_{ISB\,(ADJ)} = 3.143 \tag{3.33}$$

Table 3.12 Frequency distribution of adjusted variance for ISB

Variance adjusted $s_{ISB(ADJ)}^2$	0.44	2.94	7.11	9.44	11.11	14	20.28	Sum
Frequency (f)	1	6	1	6	1	6	6	27
Probability $(P_{ISB}^{(2)})$	1/27	6/27	1/27	6/27	1/27	6/27	6/27	1

Table 3.13 Frequency distribution of the adjusted standard deviation for ISB

Adjusted $sd_{ADJ\,(ISB)}$	0.66	1.716	2.67	3.07	3.33	3.74	4.50	Sum
Frequency (f)	1	6	1	6	1	6	6	27
Probability ($P_{ISB}^{(3)}$)	1/27	6/27	1/27	6/27	1/27	6/27	6/27	1

The MSE of the adjusted standard deviation for ISB is given by

$$\text{MSE}\left(sd_{ISB\,(ADJ)}\right) = \sum_{ISB} P_{ISB}^{(3)} \left[sd_{ISB\,(ADJ)} - \sigma_y\right]^2 = 1.539 \tag{3.34}$$

From Table 3.14, the expected value of the adjusted coefficient of variation for ISB is given by:

$$\text{E}\left(\hat{C}_{ISB\,(ADJ)}\right) = \sum_{ISB} P_{ISB}^{(4)} \hat{C}_{ISB\,(ADJ)} = 0.1944 \tag{3.35}$$

The MSE of the adjusted coefficient of variation for ISB is given by

$$\text{MSE}\left(\hat{C}_{ISB\,(ADJ)}\right) = \sum_{ISB} P_{ISB}^{(4)} \left[\hat{C}_{ISB\,(ADJ)} - C_y\right]^2 = 0.0055 \tag{3.36}$$

Thus, the relative bias using ISB for sample standard deviation is given by

$$\text{RB}\left(\hat{s}_{ISB\,(ADJ)}\right) = \frac{\text{E}\left(S_{ISB\,(ADJ)}\right) - \sigma_y}{\sigma_y} \times 100\% = -15.99\% \tag{3.37}$$

Thus, the relative bias using ISB for the sample coefficient of variation is given by

$$\text{RB}\left(\hat{C}_{ISB\,(ADJ)}\right) = \frac{\text{E}\left(\hat{C}_{ISB\,(ADJ)}\right) - C_y}{C_y} \times 100\% = -16.81\% \tag{3.38}$$

Table 3.14 Frequency distribution of the adjusted coefficient of variation for ISB

Adjusted $\hat{C}_{ISB\,(ADJ)}$	0.043	0.119	0.177	0.182	0.188	0.234	0.276	Sum
Frequency (f)	1	6	6	1	1	6	6	27
Probability ($P_{ISB}^{(4)}$)	1/27	6/27	6/27	1/27	1/27	6/27	6/27	1

Table 3.15 Comparing the variance (or MSE) of ISB and CB

	CB	ISB
Mean	4.67	1.210
Variance	98	49.40
Standard deviation	3.09	1.538
Coefficient of variation	0.0113	0.0055

When we compare the relative bias for the standard deviation of ISB in equation (3.37) with that of CB and SB found in equations (3.10) and (3.23), respectively, we see that the even though the relative bias of ISB is much less than that of CB, it is slightly higher than that of SB. This is the exact same case when we compare the relative bias for the coefficient of variation found in Equations (3.38), and (3.24) and (3.11), respectively (Tables 3.15 and 3.16).

We can easily calculate the relative efficiency of the variance of ISB with respect to the variance obtained from CB.

The percent relative efficiency of ISB over CB while considering with respect to the problem of estimation of mean is given by

$$RE(\bar{y}_{ISB}) = \frac{V(\bar{y}_{CB})}{V(\bar{y}_{ISB})} \times 100\% = 385.9\% \qquad (3.39)$$

The percent relative efficiency of ISB over CB with respect to the problem of estimation of variance is given by

$$RE\left(s_{ISB}^2\right) = \frac{V\left(s_{CB}^2\right)}{MSE\left(s_{SB}^2\right)} \times 100\% = 198.38\% \qquad (3.40)$$

The percent relative efficiency of ISB over CB with respect to the problem of estimation of standard deviation is given by

$$RE(sd_{ISB}) = \frac{MSE(sd_{CB})}{MSE(sd_{ISB(ADJ)})} \times 100\% = 200.9\% \qquad (3.41)$$

Table 3.16 Comparing the variance (or MSE) of SB and ISB

Parameters	SB	ISB
Mean	3.889	1.210
Variance	212.3	49.40
Standard deviation	3.880	1.538
Coefficient of variation	0.0132	0.0055

The percent relative efficiency of ISB over CB with respect to the problem of estimation of coefficient of variation is given by

$$\text{RE}(\hat{C}_{ISB)}) = \frac{\text{MSE}\,(\hat{C}_{CB})}{\text{MSE}\,(\hat{C}_{ISB\,(ADJ)})} \times 100\% = 205.5\% \tag{3.42}$$

Clearly, we see and conclude that ISB is more efficient than CB.

We can easily calculate the relative efficiency of ISB with respect to SB as follows.

The percent relative efficiency of ISB over SB with respect to the problem of estimation of mean is given by

$$\text{RE}(\bar{y}_{ISB}) = \frac{\text{V}\,(\bar{y}_{SB})}{\text{V}\,(\bar{y}_{ISB})} \times 100\% = 321.4\% \tag{3.43}$$

The percent relative efficiency of ISB over SB with respect to the problem of estimation of variance is given by

$$\text{RE}\left(s^2_{ISB\,(ADJ)}\right) = \frac{\text{MSE}\,(s^2_{SB})}{\text{MSE}\left(s^2_{ISB\,(ADJ)}\right)} \times 100\% = 429.7\% \tag{3.44}$$

The percent relative efficiency of ISB over SB with respect to the problem of estimation of standard deviation is given by

$$\text{RE}(sd_{ISB}) = \frac{\text{MSE}\,(sd_{SB})}{\text{MSE}\,(sd_{ISB\,(ADJ)})} \times 100\% = 252.2\% \tag{3.45}$$

The percent relative efficiency of ISB over SB with respect to the problem of estimation of coefficient of variation is given by

$$\text{RE}(\hat{C}_{ISB}) = \frac{\text{MSE}\,(\hat{C}_{SB})}{\text{MSE}\,(\hat{C}_{ISB\,(ADJ)})} \times 100\% = 240.0\% \tag{3.46}$$

Having compared the relative efficiency of ISB to that of both CB and SB we lead to the conclusion that based on our numerical illustration, ISB is more efficient.

3.5 THEORETICAL DEVELOPMENTS OF THREE TYPES OF BOOTSTRAPPING

In the following three sub sections, we discuss theoretical results for the three types of bootstrapping.

3.5.1 Conventional bootstrapping

Suppose we let s = $(y_1, y_2, \ldots \ldots \ldots y_n)$ be the initial sample with size n. We set $s_b = (y_{b_i} : i = 1, 2, \ldots, n)$ as representing the bth boot among the n^n possible boots. We have the sample mean of the bth CB as:

$$\bar{y}_{CB} = \frac{1}{n} \sum_{i \in s_u} y_{b_i} \tag{3.47}$$

Under CB for each boot, $\bar{y}_{CB}$ is the unbiased estimator of the original sample mean. The variance of $\bar{y}_{CB}$ is given by

$$V(\bar{y}_{CB}) = \frac{1}{n} s_y^2 \tag{3.48}$$

where

$$s_y^2 = (n - 1)^{-1} \sum_{i \in s} (y_i - \bar{y}_n)^2 \tag{3.49}$$

3.5.2 Sufficient bootstrapping

Let $s_b^u = (y_{i(b)}^u$ where $i(b) = 1, 2, \ldots, u(b))$ be a listing of the unique units that appear in the bth boot, that is, the bth boot contains $u(b)$ distinct units when $1 \leq u(b) \leq n$. Then an estimator of the original sample mean $\bar{y}_n$ is given by

$$\bar{y}_{SB} = \frac{1}{u} \sum_{i \in s_v} y_{i(b)}^u \tag{3.50}$$

Singh and Sedory [1] have proved the following results:

1. Under SB sampling method the mean of the samples (i.e., boots) is unbiased for the mean of the original sample that is:

$$E(\bar{y}_{SB}) = E_1 E_2 \left(\frac{1}{u} \sum_{i \in s_u} y_{i(b)}^{(u)} \right) = E_1 \left(\sum_{i \in s} (y_i) \right) = \bar{y}_n \tag{3.51}$$

where E_2 is the expected value over all possible values of u; E_1 is the expected value over all samples of the sufficient bootstrap

2. Under SB sampling method the variance of the sample means is given by

$$V(\bar{y}_{SB}) = \left[E_d\left(\frac{1}{u}\right) - \frac{1}{n} \right] s_y^2 \tag{3.52}$$

where E_d is the expected value of all possible distinct units (u) that might occur.

3.5.3 Improved sufficient bootstrapping

In the case of ISB, we introduce the following new theorems.

Theorem 5.1: Under ISB method an unbiased estimator of the original mean is given by

$$\bar{y}_{ISB} = \frac{u\bar{y}_u + (n-u)\bar{y}_n}{n} \tag{3.53}$$

where n is the size of the original sample, $\bar{y}_n$ is the mean of the values of the unique units in the current boot, and $\bar{y}_u$ is the mean of the values of the unique units in the current boot, and u is the number of these unique units in the current sample.

Proof: Let E_1 denote the expected value over all possible values of n; and E_2 denote the expected value over all possible values for u for given value of n. Taking the expected values on both sides of (3.53) we get

$$
\begin{aligned}
E(\bar{y}_{ISB}) &= E_1 E_2\left[\frac{u\bar{y}_u + (n-u)\bar{y}_n}{n} | u, n \right] \\
&= E_1\left[\frac{uE_2(\bar{y}_u) + (n-u)E_2(\bar{y}_n)}{n} | u, n \right] \\
&= E_1\left[\frac{u\bar{y}_n + (n-u)\bar{y}_n}{n} | u, n \right] \\
&= E_1(\bar{y}_n) = \bar{y}_n
\end{aligned}
$$

which proves the theorem.

Theorem 5.2: The variance of ISB estimator $\bar{y}_{ISB}$ is given by

$$V(\bar{y}_{ISB}) = \left(\frac{n-1}{n^{n+2}}\right)[(n-1)^n - (n-2)^n] s_y^2 \tag{3.54}$$

Proof: Let V_1 be the variance over all the possible values of n, and let V_2 be the variance over all possible value of u for a given value of n. Then we have

$$
\begin{aligned}
V\left(\bar{y}_{ISB}\right) &= E_1 V_2\left(\bar{y}_{ISB}|u, n\right) + V_1 E_2\left(\bar{y}_{ISB}|u, n\right) \\
&= E_1 V_2\left[\frac{u\bar{y}_u + (n-u)\bar{y}_n}{n}\right] + V_1 E_2\left[\frac{u\bar{y}_u + (n-u)\bar{y}_n}{n}|u, n\right] \\
&= E_1\left[\frac{u^2 V_2\left(\bar{y}_u\right) + (n-u)^2 V_2\left(\bar{y}_n\right) + 2u(n-u)\,cov\left(\bar{y}_n, \bar{y}_u\right)}{n^2}|n, v\right] \\
&\quad + V_1\left[\frac{uE_2\left(\bar{y}_u\right) + (n-u) E_2\left(\bar{y}_n\right)}{n}|u, n\right] \\
&= E_1\left[\left(\frac{u}{n}\right)^2\left(\frac{1}{u} - \frac{1}{n}\right)s_y^2 + 0 + 0\right] + V_1\left[\frac{u\bar{y}_n + (n-u)\bar{y}_n}{n}|u, n\right] \\
&= E_1\left[\left(\frac{u}{n^2} - \frac{u^2}{n^3}\right)s_y^2\right] + V_1\left(\bar{y}_n|n, u\right) \\
&= \frac{1}{n}E_1\left[\frac{u}{n} - \left(\frac{u}{n}\right)^2\right]s_y^2 + 0 \\
&= \frac{1}{n}E_1\left[\frac{u}{n} - \left(\frac{u}{n}\right)^2\right]s_y^2
\end{aligned}
$$

(3.55)

In (3.55), the final variance depends on the expected value of u. Now from Feller [13], Basu [14], Raj and Khamis [15] and Pathak [16], we have the following results:

$$
E_1(u) = n\left[1 - \left(\frac{n-1}{n}\right)^n\right]
$$

(3.56)

$$
E_1[u(u-1)] = E_1[u^2 - u] = E_1(u^2) - E_1(u)
$$

(3.57)

This implies that

$$
\begin{aligned}
E_1(u^2) &= E_1[u(u-1)] + E_1(u) \\
&= n(n-1)\left[1 - 2\left(\frac{n-1}{n}\right)^n + \left(\frac{n-2}{n}\right)^n\right] + n\left[1 - \left(\frac{n-1}{n}\right)^n\right]
\end{aligned}
$$

(3.58)

From (3.56) to (3.58), we have

$$
E_1\left[\frac{u}{n}\right] = 1 - \left(\frac{n-1}{n}\right)^n
$$

(3.59)

and

$$E_1\left[\frac{u^2}{n^2}\right] = \left(\frac{n-1}{n}\right)\left[1 - 2\left(\frac{n-1}{n}\right)^n + \left(\frac{n-2}{n}\right)^n\right] + \frac{1}{n}\left[1 - \left(\frac{n-1}{n}\right)^n\right] \quad (3.60)$$

On using (3.59) and (3.60) into (3.55), we have

$$\begin{aligned}
V(\bar{y}_{ISB}) &= \frac{1}{n}\left[E_1\left(\frac{u}{n}\right) - E_1\left(\frac{u^2}{n^2}\right)\right]s_y^2 \\
&= \frac{1}{n}\left[1 - \left(\frac{n-1}{n}\right)^n - \frac{1}{n}\left\{1 - \left(\frac{n-1}{n}\right)^n\right\} - \left(\frac{n-1}{n}\right)\left\{1 - 2\left(\frac{n-1}{n}\right)^n\right. \right. \\
&\quad \left. \left. + \left(\frac{n-2}{n}\right)^n\right\}\right]s_y^2 \\
&= \frac{1}{n}\left[\left\{1 - \left(\frac{n-1}{n}\right)^n\right\}\left(1 - \frac{1}{n}\right) - \left(\frac{n-1}{n}\right)\left\{1 - 2\left(\frac{n-1}{n}\right)^n\right. \right. \\
&\quad \left. \left. + \left(\frac{n-2}{n}\right)^n\right\}\right]s_y^2 \\
&= \left(\frac{n-1}{n^2}\right)\left[1 - \left(\frac{n-1}{n}\right)^n - \left\{1 - 2\left(\frac{n-1}{n}\right)^n + \left(\frac{n-2}{n}\right)^n\right\}\right]s_y^2 \\
&= \left(\frac{n-1}{n^2}\right)\left[\left(\frac{n-1}{n}\right)^n - \left(\frac{n-2}{n}\right)^n\right]s_y^2 \\
&= \left(\frac{n-1}{n^{n+2}}\right)\left[(n-1)^n - (n-2)^n\right]s_y^2
\end{aligned}$$

$$(3.61)$$

which proves the theorem.

Theorem 5.3: An adjusted estimator of variance is given by

$$s_{ISB\,(ADJ)}^2 = \frac{(n-1)\,S_{ISB}^2 - \frac{(n-u)\,u^2\,(\bar{y}_n - \bar{y}_u)^2}{n^2}}{u} \quad (3.62)$$

Proof: We have

$$s_{ISB}^2 = \frac{\sum_{i=1}^{u}\left[y_i - \frac{u\bar{y}_u + (n-u)\,\bar{y}_n}{n}\right]^2 + \sum_{i=u+1}^{n}\left[\bar{y}_n - \frac{u\bar{y}_u + (n-u)\,\bar{y}_n}{n}\right]^2}{(n-1)}$$

So we have

$$
\begin{aligned}
(n-1)s_{ISB}^2 &= \sum_{i=1}^{u}\left[y_i - \frac{u\bar{y}_u + (n-u)\bar{y}_n}{n}\right]^2 + \sum_{i=u+1}^{n}\left[\bar{y}_n - \frac{u\bar{y}_u + (n-u)\,\bar{y}_n}{n}\right]^2 \\
&= \sum_{i=1}^{u}\left[y_i - \frac{u\bar{y}_u + (n-u)\,\bar{y}_n}{n}\right]^2 + \sum_{i=u+1}^{n}\left[\frac{n\bar{y}_n - u\bar{y}_u - (n-u)\,\bar{y}_n}{n}\right]^2 \\
&= \sum_{i=1}^{u}\left[y_i - \frac{u\bar{y}_u + (n-u)\,\bar{y}_n}{n}\right]^2 + \sum_{i=u+1}^{n}\left[\frac{n\bar{y}_n - u\bar{y}_u - n\bar{y}_n + u\bar{y}_n}{n}\right]^2 \\
&= \sum_{i=1}^{u}\left[y_i - \frac{u\bar{y}_u + (n-u)\,\bar{y}_n}{n}\right]^2 + \frac{(n-u)}{n^2}\left[u\left(\bar{y}_n - \bar{y}_u\right)\right]^2 \\
&= \sum_{i=1}^{u}\left[y_i - \frac{u\bar{y}_u + (n-u)\,\bar{y}_n}{n}\right]^2 + \frac{(n-u)u^2}{n^2}\left[\left(\bar{y}_n - \bar{y}_u\right)\right]^2
\end{aligned}
$$

This implies

$$
\sum_{i=1}^{u}\left[y_i - \frac{u\bar{y}_u + (n-u)\,\bar{y}_n}{n}\right]^2 = (n-1)s_{ISB}^2 - \frac{(n-u)u^2}{n^2}\left[\left(\bar{y}_n - \bar{y}_u\right)\right]^2
$$

Because $u \geq 1$, so to avoid division by zero, we suggest an adjusted estimator for estimator of variance written as

$$
s_{ISB\,(ADJ)}^2 = \frac{\sum_{i=1}^{u}\left[y_i - \frac{u\bar{y}_u + (n-u)\,\bar{y}_n}{n}\right]^2}{u}
$$

Or equivalently

$$
s_{ISB\,(ADJ)}^2 = \frac{(n-1)S_{ISB}^2 - \frac{(n-u)u^2(\bar{y}_n - \bar{y}_u)^2}{n^2}}{u}
$$

In the next section, we do an analytical comparison of ISB with respect to CB.

3.6 ANALYTICAL COMPARISON

We know that the variance of the conventional bootstrap is given by

$$
V(\bar{y}_{CB}) = \frac{s_y^2}{n} \tag{3.63}
$$

From (3.54) and (3.63), we define the relative efficiency (RE(ISB)) of the ISB over the CB as

$$
\begin{aligned}
\mathrm{RE}\left(\bar{y}_{ISB}\right) &= \frac{\mathrm{Var}\left(\bar{y}_{CB}\right)}{\mathrm{Var}\left(\bar{y}_{ISB}\right)} \times 100\% \\[2mm]
&= \frac{\frac{s_y^2}{n}}{\left(\frac{n-1}{n^{n+2}}\right)\left[(n-1)^n - (n-2)^n\right]s_y^2} \times 100\% \\[2mm]
&= \frac{n^{n+2}}{n(n-1)\left[(n-1)^n - (n-2)^n\right]} \times 100\% \\[2mm]
&= \frac{n^{n+1}}{(n-1)\left[(n-1)^n - (n-2)^n\right]} \times 100\%
\end{aligned}
\tag{3.64}
$$

The values of RE $\left(\bar{y}_{ISB}\right)$ in (3.64) for even sample sizes ranging from 2 to 140 are given in Table 3.17.

Table 3.17 The RE (%) for ISB with various sample sizes

n	RE $\left(\bar{y}_{ISB}\right)$	n	RE $\left(\bar{y}_{ISB}\right)$	n	RE $\left(\bar{y}_{ISB}\right)$
2	800.00	50	282.26	98	277.08
4	451.10	52	281.84	100	276.97
6	378.14	54	281.46	102	276.87
8	347.05	56	281.11	104	276.77
10	329.95	58	280.79	106	276.68
12	319.17	60	280.48	108	276.59
14	311.75	62	280.19	110	276.50
16	306.34	64	279.93	112	276.42
18	302.21	66	279.68	114	276.33
20	298.97	68	279.44	116	276.26
22	296.35	70	279.22	118	276.18
24	294.19	72	279.01	120	276.11
26	292.38	74	278.81	122	276.04
28	290.85	76	278.63	124	275.97
30	289.52	78	278.45	126	275.90
32	288.37	80	278.28	128	275.84
34	287.36	82	278.12	130	275.77
36	286.46	84	277.97	132	275.71
38	285.67	86	277.82	134	275.66
40	284.95	88	277.69	136	275.60
42	284.31	90	277.55	137	275.57
44	283.72	92	277.43	138	275.54
46	283.19	94	277.31	139	275.52
48	282.70	96	277.19	140	275.49

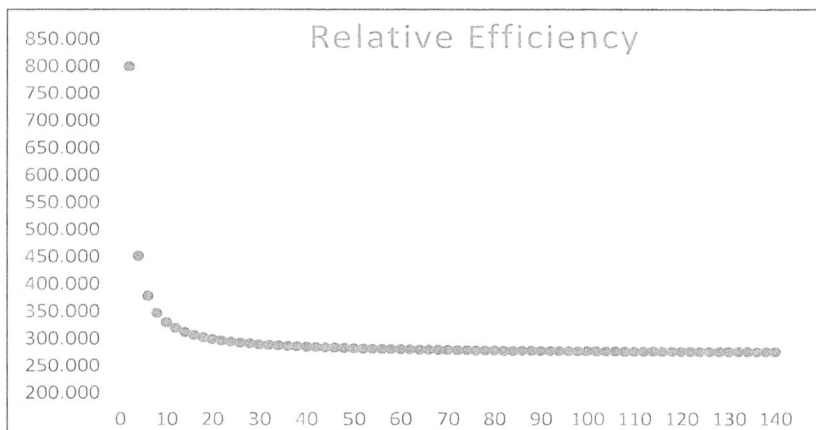

Figure 3.1 Visualization of RE(ISB) for different sample sizes.

To have a different view of the value of the percent RE $(\bar{y}_{ISB})$ for different sample sizes, we have displayed the results graphically in Figure 3.1. One can easily see that in the case of small sample sizes ISB performs better than CB. For large sample sizes, the value of RE $(\bar{y}_{ISB})$ decreases substantially, but still remains more than 250%. It is worth noting that Excel and SAS are unable to compute the value of RE $(\bar{y}_{ISB})$ for a sample size greater than 140. We also remark that the most dramatic decreases in relative efficiency occur for smaller sample sizes.

With the results shown in Table 3.17 serving as encouragement, we investigate further the properties of ISB by means of simulation study.

3.7 SIMULATION STUDY

In the simulation study, we will look at the performance of the three bootstrapping methods in regard to estimating the mean, variance, standard deviation and coefficient of variation. Our variables will come from a population following a beta distribution $B(\alpha, \beta)$, the particular ones considered corresponding to $\alpha = 2.5$ *and* 4.5, and $\beta = 2.5$ *and* 4.5. Thus, there are four possibilities. Recall that the values of the parameters to be estimated are given by:

$$\text{Population Mean: } \mu_y = \frac{\alpha}{\alpha + \beta} = \theta_1 \tag{3.65}$$

$$\text{Population variance: } \sigma_y^2 = \frac{\alpha\beta}{(\alpha + \beta)^2 (\alpha + \beta + 1)} = \theta_2 \tag{3.66}$$

Population Standard: $\sigma_y = \sqrt{\sigma_y^2} = \theta_3$ (3.67)

and

Coefficient of variation: $C_y = \dfrac{\sigma_y}{\overline{Y}} \times 100\% = \theta_4$ (3.68)

For various values of n (namely 10 to 100 with a step of 5), we generate a random sample of size n from the current population. For each of these samples we generate $nboots$ = 5,000 bootstrap samples (boots), each being an SRS with replacement sample of size n taken from the original. This is done by means of the R function runif(n,1,n). For each boot, we process it in one of the following ways: (1) as a CB sample where it is left as is; (2) as a SB sample where it has repeated units removed; and (3) as an ISB where repeated units are replaced by the mean of the original sample.

We obtain three estimates of each parameter of interest as developed in the previous sections. For example, concerning the mean (j = 1) we have $\hat{\theta}_{1(CB)}$ as the sample mean $\overline{y}_n$ of the CB-boot (a size n sample); $\hat{\theta}_{1\,(SB)}$ as the sample mean $\overline{y}_u$ of the SB-boot (a size u sample with u being the number of distinct units); and $\hat{\theta}_{1\,(ISB)}$ being the sample mean $\overline{y}_{ISB}$ (adjusted) of the ISB-boot. Similarly, we have $\hat{\theta}_{j\,(CB)|b}, \hat{\theta}_{j\,(SB)|b}$ and $\hat{\theta}_{j\,(ISB)|b}$ for j = 1, 2, 3, 4; b = 1, 2, 3, ...,$nboots$, where j = 2 corresponds to variance, j = 3 to standard deviation and j = 4 to coefficient of variation. To gauge the performance of the estimators, we compute the following for each population and sample sizes.

Relative Biases (RB) are computed as:

$$RB(\hat{\theta}_j)_{CB} = \dfrac{\frac{1}{nboots} \sum_{b=1}^{nboots} \hat{\theta}_{j(CB)|b} - \theta_j}{\theta_j} \times 100\%$$ (3.69)

$$RB(\hat{\theta}_j)_{SB} = \dfrac{\frac{1}{nboots} \sum_{b=1}^{nboots} \hat{\theta}_{j(SB)|b} - \theta_j}{\theta_j} \times 100\%$$ (3.70)

$$RB(\hat{\theta}_j)_{ISB} = \dfrac{\frac{1}{nboots} \sum_{b=1}^{nboots} \hat{\theta}_{j(ISB)|b} - \theta_j}{\theta_j} \times 100\%$$ (3.71)

Mean Square Errors are computed as:

$$MSE(\hat{\theta}_j)_{CB} = \dfrac{1}{nboots} \sum_{b=1}^{nboots} \left(\hat{\theta}_{j(CB)|b} - \theta_j \right)^2$$ (3.72)

$$\text{MSE}(\hat{\theta}_j)_{SB} = \frac{1}{nboots} \sum_{b=1}^{nboots} \left(\hat{\theta}_{j_{(SB)|b}} - \theta_j\right)^2 \tag{3.73}$$

and

$$\text{MSE}(\hat{\theta}_j)_{ISB} = \frac{1}{nboots} \sum_{b=1}^{nboots} \left(\hat{\theta}_{j_{(ISB)|b}} - \theta_j\right)^2 \tag{3.74}$$

Relative Efficiency values are computed as:

$$\text{RE}\left[\left(\hat{\theta}_j\right)_{CB}, \left(\hat{\theta}_j\right)_{SB}\right] = \frac{\text{MSE}(\hat{\theta}_j)_{CB}}{\text{MSE}(\hat{\theta}_j)_{SB}} \times 100\% \tag{3.75}$$

$$\text{RE}\left[\left(\hat{\theta}_j\right)_{CB}, \left(\hat{\theta}_j\right)_{ISB}\right] = \frac{\text{MSE}(\hat{\theta}_j)_{CB}}{\text{MSE}(\hat{\theta}_j)_{ISB}} \times 100\% \tag{3.76}$$

and

$$\text{RE}\left[\left(\hat{\theta}_j\right)_{SB}, \left(\hat{\theta}_j\right)_{ISB}\right] = \frac{\text{MSE}(\hat{\theta}_j)_{SB}}{\text{MSE}(\hat{\theta}_j)_{ISB}} \times 100\% \tag{3.77}$$

The results of the simulation study for relative efficiencies are given in Tables 3.18–3.20. Each figure is the result of the four values obtained from the four beta populations. From Table 3.18, for sample size 10, the average value of RE(m,1), that is RE[$(\hat{\theta}_1)_{CB}$, $(\hat{\theta}_1)_{SB}$], is 158.95% with a standard deviation of 1.24%, the minimum value is 158.20% and the maximum value is 160.80%; the average value of RE(m,2), that is, RE[$(\hat{\theta}_1)_{CB}$, $(\hat{\theta}_1)_{ISB}$], is 423.08% with a standard deviation of 0.395%, the minimum value is 422.60% and the maximum value is 423.40%; the average value of RE (m,3), that is, RE[$(\hat{\theta}_1)_{SB}$, $(\hat{\theta}_1)_{ISB}$], is 266.13% with a standard deviation of 2.22%, the minimum value is 262.80% and the maximum value is 267.30%; for four combinations of α and β. Interpretation of the results for other sample sizes is similar. The table includes results for other parameters that is, RE(s^2,j), RE(sd,j) and RE(cv,j). Likewise, the rest of the table can be interpreted.

In Table 3.19, we have for each of the two beta populations with $\alpha = 2.5$ averaged the RE values of a parameter and a particular combination as listed over 19 sample sizes. For $\beta = 2.5$, the average RE(m,1) is 168.70% with a standard deviation of 4.66%, minimum value of 160.78% and maximum value of 176.66%; the average RE(m,2) is 427.91% with a

Table 3.18 Descriptive statistics: RE(m,1), RE(m,2), RE(m,3), RE(s²,1), …, 2), RE(cv,3)

n	Freq	Mean	StDev	Min	Max	Mean	StDev	Min	Max	Mean	StDev	Min	Max
		RE(m,1)				RE(mean) RE(m,2)				RE(m,3)			
10	4	158.95	1.24	158.20	160.80	423.08	0.395	422.60	423.40	266.13	2.22	262.80	267.30
15	4	162.20	2.08	160.40	164.10	420.78	4.83	416.60	425.20	259.45	0.404	259.10	259.80
20	4	164.38	1.59	163.00	165.80	422.45	2.25	420.50	424.50	257.02	1.13	256.00	258.00
25	4	162.58	0.263	162.30	162.80	415.85	0.640	415.20	416.40	255.82	0.0500	255.80	255.90
30	4	163.95	1.57	162.60	165.50	416.98	2.78	414.60	419.90	254.27	0.723	253.60	254.90
35	4	171.93	0.960	170.50	172.60	434.15	2.71	430.10	435.70	252.53	0.435	252.10	252.90
40	4	165.72	1.42	164.50	167.10	419.33	5.58	414.50	424.40	253.03	1.18	252.00	254.10
45	4	166.65	3.53	163.40	169.70	419.77	9.41	410.80	427.90	251.88	0.330	251.40	252.10
50	4	169.68	4.68	167.30	176.70	428.68	11.82	422.70	446.40	252.65	0.100	252.50	252.70
55	4	171.68	5.23	167.10	176.20	435.25	12.99	423.90	446.50	253.50	0.115	253.40	253.60
60	4	170.80	4.42	166.30	176.90	432.60	9.44	421.90	444.90	253.28	1.20	251.50	254.00
65	4	170.53	1.12	169.90	172.20	430.73	3.99	428.60	436.70	252.55	0.714	252.10	253.60
70	4	174.60	3.88	171.70	179.90	439.42	9.92	432.10	453.10	251.70	0.163	251.50	251.90
75	4	165.82	3.61	162.90	170.30	416.88	10.39	408.50	429.90	251.38	0.759	250.80	252.40
80	4	169.78	2.30	168.30	173.20	427.27	6.09	423.90	436.40	251.72	0.263	251.50	252.00
85	4	169.78	4.32	167.00	176.10	427.90	10.06	421.80	442.80	252.00	0.583	251.40	252.50
90	4	173.15	2.53	170.40	175.30	436.05	6.88	428.60	441.90	251.85	0.289	251.60	252.10
95	4	168.28	1.92	165.80	169.80	422.02	5.41	415.60	426.50	250.72	0.479	250.10	251.10
100	4	171.48	0.885	171.00	172.80	431.75	2.20	430.40	435.00	251.80	0.141	251.70	252.00

(Continued)

Table 3.18 (Continued) Descriptive statistics: RE(m,1), RE(m,2), RE(m,3), RE(s²,1), ..., 2), RE(cv,3)

n	Freq	RE(s²,1)				RE(variance) RE(s²,2)				RE(s²,3)			
		Mean	StDev	Min	Max	Mean	StDev	Min	Max	Mean	StDev	Min	Max
10	4	98.025	0.395	97.700	98.500	163.88	4.56	157.20	166.80	167.15	5.04	160.00	170.70
15	4	121.65	0.866	120.90	122.40	162.28	1.14	161.40	163.80	133.35	1.84	131.80	135.40
20	4	130.07	0.960	129.10	130.90	164.10	3.02	161.10	166.70	126.13	1.41	124.50	127.30
25	4	137.20	1.16	136.10	138.20	162.95	1.45	161.50	164.20	118.83	0.222	118.50	119.00
30	4	140.95	0.252	140.60	141.20	163.35	0.412	162.90	163.90	115.90	0.503	115.40	116.60
35	4	148.20	2.34	145.30	150.10	169.57	2.79	166.30	171.90	114.40	0.141	114.20	114.50
40	4	152.85	2.14	151.00	154.70	168.65	2.60	166.20	170.90	110.35	0.191	110.10	110.50
45	4	157.40	2.87	154.10	159.80	175.85	3.17	172.20	178.50	111.73	0.0500	111.70	111.80
50	4	151.55	2.72	149.20	154.10	164.60	4.04	161.10	168.10	108.60	0.712	108.00	109.40
55	4	157.10	1.16	156.10	158.20	169.88	0.723	169.30	170.80	108.15	0.436	107.60	108.50
60	4	156.80	3.37	154.40	161.80	168.13	4.19	165.70	174.40	107.23	0.427	106.90	107.80
65	4	162.68	2.01	159.70	163.90	174.02	2.07	171.00	175.70	107.00	0.523	106.60	107.70
70	4	162.63	3.08	158.00	164.30	172.27	2.56	168.70	174.80	105.93	0.732	105.30	106.70
75	4	159.50	1.47	157.30	160.30	169.00	1.62	166.60	170.00	105.93	0.0957	105.80	106.00
80	4	162.82	1.86	160.80	164.40	173.47	3.18	170.20	176.20	106.58	0.723	105.90	107.20
85	4	160.20	2.38	157.10	162.90	169.35	2.13	166.60	171.80	105.68	0.222	105.50	106.00
90	4	161.43	3.92	158.10	165.70	168.85	3.87	165.50	172.30	104.63	0.457	104.00	105.10
95	4	161.68	2.92	159.60	165.80	168.93	3.82	166.30	174.40	104.50	0.476	104.20	105.20
100	4	164.10	0.952	163.50	165.50	170.70	0.800	170.30	171.90	104.00	0.115	103.90	104.10

RE(standard deviation)

		RE(sd,1)				RE(sd,2)				RE(sd,3)			
10	4	113.16	0.79	111.96	113.58	152.65	7.80	141.42	157.95	134.88	6.04	126.31	139.12
15	4	127.95	2.86	125.48	130.68	146.00	5.40	141.38	151.64	114.08	1.69	112.67	116.04
20	4	136.99	0.37	136.50	137.28	154.21	2.17	151.53	155.97	112.57	1.28	111.01	113.61
25	4	140.85	0.42	140.25	141.16	152.42	0.62	151.49	152.81	108.21	0.20	108.01	108.50
30	4	142.14	0.99	141.44	143.56	152.50	1.57	151.63	154.86	107.29	0.48	106.69	107.87
35	4	151.83	1.12	150.34	152.69	162.77	1.32	161.14	163.84	107.21	0.12	107.03	107.30
40	4	155.83	2.96	152.97	158.39	162.45	3.28	159.16	165.28	104.25	0.14	104.05	104.35
45	4	161.24	3.69	157.12	164.36	171.35	3.93	166.88	174.65	106.27	0.04	106.21	106.33
50	4	151.21	3.18	148.54	154.83	157.01	4.21	153.41	161.39	103.83	0.64	103.28	104.51
55	4	158.66	2.85	156.20	161.49	164.63	2.47	162.55	167.41	103.77	0.38	103.27	104.07
60	4	157.87	3.61	155.77	163.24	162.95	4.35	160.30	169.39	103.21	0.41	102.90	103.77
65	4	161.16	2.03	159.73	164.17	166.51	2.77	165.09	170.67	103.31	0.47	102.96	103.96
70	4	162.19	2.24	159.13	164.51	166.34	2.17	164.38	169.45	102.56	0.69	101.97	103.30
75	4	161.17	2.26	157.80	162.45	165.68	2.40	162.12	167.13	102.79	0.10	102.67	102.88
80	4	165.19	2.27	162.63	167.11	171.13	3.56	167.33	174.17	103.58	0.73	102.89	104.22
85	4	161.42	2.34	159.22	164.73	166.17	2.08	164.39	169.17	102.94	0.22	102.70	103.25
90	4	161.36	3.81	158.26	166.06	164.64	3.47	161.68	168.36	102.04	0.45	101.38	102.44
95	4	162.59	3.18	160.37	167.11	165.84	4.01	163.10	171.60	101.99	0.47	101.70	102.69
100	4	164.99	1.44	163.87	166.89	167.77	1.27	166.84	169.52	101.69	0.14	101.55	101.81

(Continued)

Table 3.18 (Continued) Descriptive statistics: RE(m,1), RE(m,2), RE(m,3), RE(s²,1), ..., 2), RE(cv,3)

n	Freq	RE(cv,1)				RE(coefficient of variation) RE(cv,2)				RE(cv,3)			
		Mean	StDev	Min	Max	Mean	StDev	Min	Max	Mean	StDev	Min	Max
10	4	119.82	3.31	115.37	122.43	183.28	10.13	174.79	196.93	152.91	5.68	147.97	161.09
15	4	138.73	3.90	134.38	143.72	175.88	19.09	154.02	200.58	126.65	11.38	111.97	139.56
20	4	145.24	2.86	141.73	147.58	199.65	16.97	182.53	216.82	137.34	9.09	128.79	146.93
25	4	148.32	1.27	146.95	149.94	185.93	10.18	175.16	199.69	125.33	6.02	118.51	133.18
30	4	149.33	2.66	146.61	152.67	191.98	14.70	181.68	213.66	128.47	7.67	123.92	139.94
35	4	155.76	3.14	152.28	159.88	190.07	13.75	175.24	202.61	121.97	7.44	113.01	130.01
40	4	159.04	1.34	157.28	160.51	192.49	10.74	180.44	202.50	121.04	6.90	112.42	127.36
45	4	163.34	3.35	158.32	165.11	222.29	16.54	204.02	240.56	136.02	8.35	128.87	145.88
50	4	153.93	2.80	151.47	157.46	168.43	11.13	158.06	179.04	109.42	6.91	103.04	116.86
55	4	162.44	1.13	161.38	163.97	206.17	11.47	196.37	222.73	126.92	6.92	121.34	137.01
60	4	161.69	3.84	159.20	167.33	186.97	13.70	178.21	207.07	115.54	5.86	110.79	123.75
65	4	163.67	3.94	160.53	168.97	193.12	19.35	172.49	219.25	117.85	9.19	107.45	129.76
70	4	165.25	3.39	160.51	168.43	191.40	13.56	174.81	206.04	115.74	5.98	108.91	122.33
75	4	162.21	2.74	158.18	164.22	195.18	14.67	176.14	208.22	120.25	7.20	111.36	126.80
80	4	167.03	0.83	166.48	168.25	226.62	19.22	203.98	248.34	135.65	10.92	122.53	147.60
85	4	164.52	4.51	159.58	169.67	198.33	15.21	184.81	219.62	120.47	6.86	114.45	129.44
90	4	164.26	2.19	161.71	166.10	193.73	9.11	187.23	206.90	117.93	4.95	112.74	124.56
95	4	165.73	3.89	162.61	170.89	195.13	14.50	184.87	216.27	117.67	6.64	110.99	126.55
100	4	166.78	1.35	164.97	168.23	208.99	10.87	198.10	222.43	125.31	6.35	118.68	132.21

Table 3.19 Descriptive statistics: RE(m,1), RE(m,2), RE(m,3), RE(s²,1), ..., RE(cv,3) (Results for α = 2.5)

β	Freq	Mean	StDev	Min	Max	Mean	StDev	Min	Max	Mean	StDev	Min	Max
		RE(m,1)				**RE(mean)** / **RE(m,2)**				**RE(m,3)**			
2.5	19	168.70	4.66	160.78	176.66	427.91	9.31	412.47	446.39	253.70	3.02	250.13	262.84
4.5	19	167.56	4.97	158.45	176.19	425.52	9.67	408.48	446.47	254.02	3.98	250.78	267.25
		RE(s²,1)				**RE(variance)** / **RE(s²,2)**				**RE(s²,3)**			
2.5	19	150.27	18.06	98.20	165.80	168.71	5.06	157.20	175.70	113.79	13.81	103.90	160.00
4.5	19	149.99	17.36	97.70	164.40	168.72	4.81	161.10	178.50	114.12	15.74	104.10	170.70
		RE(sd,1)				**RE(standard deviation)** / **RE(sd,2)**				**RE(sd,3)**			
2.5	19	153.61	14.73	111.96	167.11	162.57	8.22	141.42	171.6	106.37	6.10	101.38	126.31
4.5	19	152.14	14.36	113.54	167.11	161.53	8.00	141.38	174.65	106.82	8.55	101.70	139.12
		RE(cv,1)				**RE(coefficient of variation)** / **RE(cv,2)**				**RE(cv,3)**			
2.5	19	159.33	12.07	122.43	170.89	209.84	14.10	179.04	234.96	132.09	8.95	113.70	151.07
4.5	19	156.10	11.53	122.25	168.25	193.99	24.25	154.02	248.34	124.61	15.64	103.04	161.09

Table 3.20 Descriptive statistics: RE(m,1), RE(m,2), RE(m,3), RE(s²,1), ..., RE(cv,3) (Results for α = 4.5)

RE(mean)

β	N	RE(m,1)				RE(m,2)				RE(m,3)			
		Mean	StDev	Min	Max	Mean	StDev	Min	Max	Mean	StDev	Min	Max
2.5	19	167.56	4.97	158.45	176.19	425.52	9.67	408.48	446.47	254.02	3.98	250.78	267.25
4.5	19	168.16	5.08	158.17	179.86	426.5	10.13	410.79	453.08	253.7	3.91	250.61	267.35

RE(variance)

β	N	RE(s²,1)				RE(s²,2)				RE(s²,3)			
		Mean	StDev	Min	Max	Mean	StDev	Min	Max	Mean	StDev	Min	Max
2.5	19	149.99	17.36	97.70	164.40	168.72	4.81	161.10	178.50	114.12	15.74	104.10	170.70
4.5	19	149.07	17.27	98.50	164.30	167.51	3.93	161.10	174.80	113.97	15.05	103.90	167.20

RE(sd)

β	N	RE(sd,1)				RE(sd,2)				RE(sd,3)			
		Mean	StDev	Min	Max	Mean	StDev	Min	Max	Mean	StDev	Min	Max
2.5	19	152.14	14.36	113.54	167.11	161.53	8	141.38	174.65	106.82	8.55	101.7	139.12
4.5	19	152.17	13.81	113.58	165.32	161.31	6.74	149.61	169.45	106.6	7.66	101.55	134.96

RE(cv)

β	N	RE(cv,1)				RE(cv,2)				RE(cv,3)			
		Mean	StDev	Min	Max	Mean	StDev	Min	Max	Mean	StDev	Min	Max
2.5	19	155.81	13.39	115.37	166.92	189.60	11.12	173.51	213.54	122.23	8.68	112.42	151.5
4.5	19	155.52	11.88	119.24	167.00	186.71	13.41	158.06	219.21	120.53	10.24	104.06	147.97

standard deviation of 9.31%, minimum value of 412.47% and maximum value of 446.39%; the average RE(m,3) is 253.7% with a standard deviation of 3.02%, minimum value of 250.13% and maximum value of 262.84%. Likewise, the rest of the table can be interpreted.

In Table 3.20, for α = 4.5 there are 19 sample sizes between 10 and 100 with a skip of 5. For and β = 2.5, the average RE(m,1) is 167.56% with a standard deviation of 4.97%, minimum value of 158.45% and maximum value of 176.19%; the average RE(m,2) is 425.52% with a standard deviation of 9.67%, minimum value of 408.48% and maximum value of 446.47%; the average RE(m,3) is 254.02% with a standard deviation of 3.98%, minimum value of 250.78% and maximum value of 267.25%. Likewise, the rest of the table can be interpreted.

In Figure 3.2 each panel represents a scatter plot of relative efficiency values versus sample size for each of the four populations. Each panel corresponds to a particular parameter and comparison.

In Figure 3.3 each panel corresponds to a particular parameter and comparison. Of four box plots of the distribution of the relative efficiency values versus sample sizes, each box plot corresponds to 19 observations for one of the four populations.

In the following tables and graphics, we have presented results on relative bias in the same way we did for relative efficiency. For each parameter comparison combinations there are 19 values of relative biases obtained from each of the four populations. We remind the readers that each of the estimators of the means is unbiased.

From Table 3.21, for sample size 10, the average four values of RB (m,1) is 0.0715% with a standard deviation of 0.156%, the minimum value is –0.135% and the maximum value is 0.028%; the average

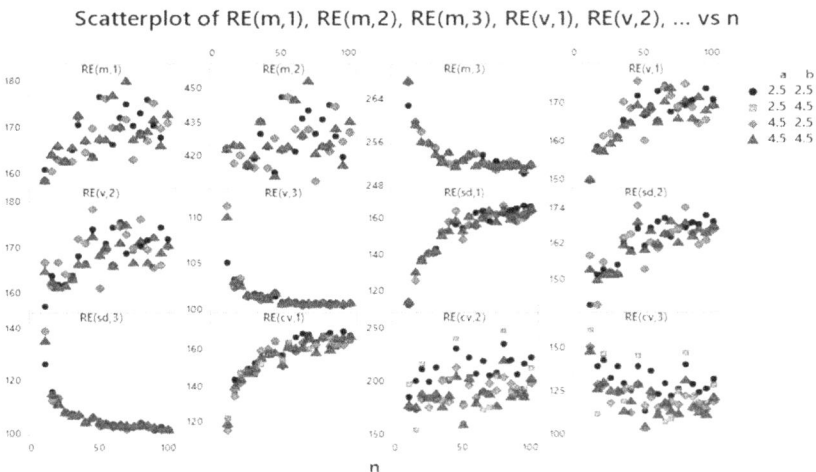

Figure 3.2 RE vs sample size for mean, variance, standard deviation and coefficient of variation.

Boxplot of RE(m,1), RE(m,2), RE(m,3), RE(v,1), RE(v,2), RE(v,3), ...

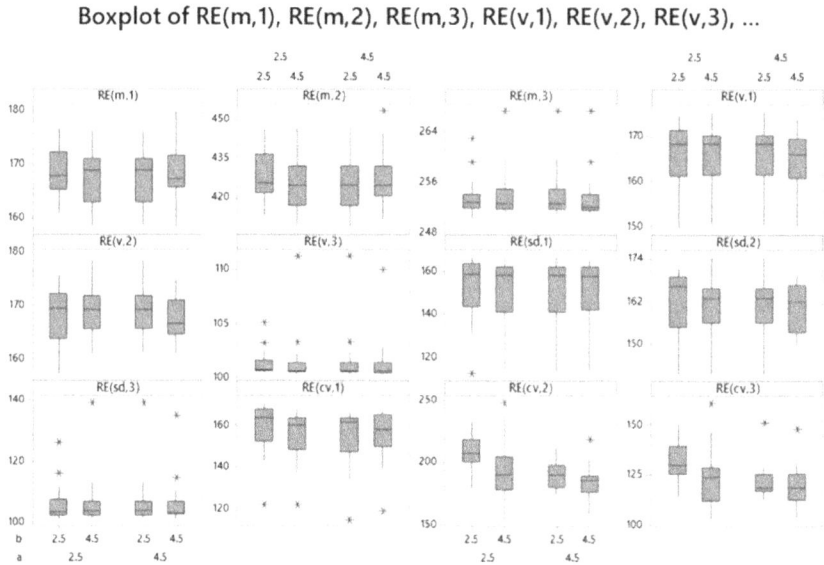

Figure 3.3 Box plots of RE vs different pairs of α and β for mean, variance, standard deviation and coefficient of variation.

value of RB(m,2) is 0.028% with a standard deviation of 0.0835%, the minimum value is –0.069% and the maximum value is 0.115%; the average value of RB(m,3) is 0.0168% with a standard deviation of 0.0465%, the minimum value is –0.039% and the maximum value is 0.065%.

In Table 3.22, for α = 2.5 and β = 2.5, there are 19 sample sizes between 10 to 100 with a skip of 5. The average RB(m,1) is –0.0103% with a standard deviation of 0.0779%, a minimum value of –0.167% and maximum value of 0.0687%; the average RB(m,2) is –0.0068% with a standard deviation of 0.0687%, minimum value of –0.136% and maximum value of 0.106%; the average RB(m,3) is –0.00458% with a standard deviation of 0.042%, minimum value of –0.084% and maximum value of 0.061. Likewise, the rest of the table can be interpreted.

In Table 3.23, for α = 4.5 and β = 2.5, there are 19 sample sizes between 10 to 100 with a skip of 5. The average RB(m,1) is –0.0164% with a standard deviation of 0.0622%, a minimum value of –0.149% and maximum value of 0.07%; the average RB(m,2) is –0.0146% with a standard deviation of 0.0513%, minimum value of –0.127% and maximum value of 0.053%; the average RB(m,3) is –0.00753% with a standard deviation of 0.03084%, minimum value of –0.067% and maximum value of 0.033%. Likewise, the rest of the table can be interpreted.

Table 3.21 Descriptive statistics: RB(m,1), RB(m,2), RB(m,3), RB(s²,1), ..., RB(cv,3)

n	Freq	RB(m,1)				RB(mean) RB(m,2)				RB(m,3)			
		Mean	StDev	Min	Max	Mean	StDev	Min	Max	Mean	StDev	Min	Max
10	4	0.0715	0.156	-0.135	0.225	0.028	0.0835	-0.069	0.115	0.0168	0.0465	-0.039	0.065
15	4	-0.046	0.0501	-0.091	0.025	-0.0735	0.0843	-0.136	0.051	-0.0458	0.0532	-0.084	0.033
20	4	0.009	0.1611	-0.149	0.234	-0.0412	0.1604	-0.129	0.199	-0.0243	0.087	-0.074	0.106
25	4	0.0695	0.1126	-0.096	0.154	0.0602	0.0975	-0.083	0.133	0.0345	0.0561	-0.048	0.076
30	4	-0.09	0.108	-0.167	0.07	-0.018	0.0288	-0.045	0.019	-0.0125	0.01859	-0.031	0.01
35	4	0.0118	0.0314	-0.022	0.039	-0.005	0.0313	-0.044	0.025	-0.0055	0.0248	-0.034	0.019
40	4	-0.023	0.0557	-0.065	0.059	0.0085	0.0661	-0.063	0.097	0.0063	0.0441	-0.042	0.065
45	4	-0.0078	0.059	-0.078	0.044	-0.0058	0.0334	-0.049	0.028	-0.004	0.0223	-0.033	0.019
50	4	0.0238	0.0605	-0.031	0.11	0.0215	0.0291	-0.019	0.045	0.01275	0.01603	-0.01	0.024
55	4	-0.0513	0.1156	-0.135	0.118	-0.035	0.1014	-0.101	0.116	-0.023	0.0646	-0.066	0.073
60	4	-0.0227	0.0462	-0.067	0.04	-0.02	0.0524	-0.068	0.041	-0.0145	0.0342	-0.044	0.027
65	4	0.0143	0.0752	-0.051	0.113	0.0227	0.0605	-0.046	0.101	0.0153	0.0386	-0.029	0.065
70	4	0.0025	0.0347	-0.044	0.031	-0.0093	0.0467	-0.058	0.036	-0.0063	0.0299	-0.038	0.024
75	4	0.025	0.0569	-0.037	0.101	0.0142	0.0248	-0.001	0.051	0.008	0.01481	-0.002	0.03
80	4	-0.0153	0.0265	-0.051	0.012	0.0227	0.0288	-0.02	0.041	0.01475	0.01945	-0.014	0.028
85	4	-0.0258	0.0685	-0.088	0.072	-0.0262	0.0646	-0.103	0.055	-0.016	0.0388	-0.062	0.033
90	4	0.0215	0.0399	-0.012	0.078	0.0018	0.0219	-0.024	0.029	0.0005	0.0125	-0.015	0.015
95	4	-0.015	0.0643	-0.082	0.042	0.0095	0.0559	-0.059	0.073	0.006	0.0365	-0.039	0.047
100	4	0.0255	0.0575	-0.057	0.075	0.0213	0.0751	-0.091	0.067	0.0142	0.0471	-0.056	0.044

(Continued)

Table 3.21 (Continued) Descriptive statistics: RB(m,1), RB(m,2), RB(m,3), RB(s²,1), …), RB(cv,3)

n	Freq	RB(s²,1)				RB(variance) RB(s²,2)				RB(s²,3)			
		Mean	StDev	Min	Max	Mean	StDev	Min	Max	Mean	StDev	Min	Max
10	4	−0.13	0.266	−0.288	0.268	11.060	0.188	10.940	11.340	−5.3335	0.1029	−5.407	−5.181
15	4	−0.1018	0.0397	−0.136	−0.064	7.0725	0.0838	7.0000	7.1500	−3.5553	0.0726	−3.618	−3.488
20	4	0.0092	0.1537	−0.137	0.142	5.2125	0.1372	5.0700	5.3300	−2.63	0.1388	−2.773	−2.511
25	4	0.2647	0.117	0.158	0.366	4.3525	0.0665	4.2900	4.4100	−1.8805	0.0632	−1.942	−1.826
30	4	0.07025	0.00954	0.062	0.079	3.5300	0.0808	3.4600	3.6000	−1.6325	0.0826	−1.707	−1.561
35	4	0.209	0.325	−0.142	0.486	3.170	0.236	2.910	3.370	−1.242	0.222	−1.481	−1.053
40	4	−0.297	0.301	−0.557	−0.036	2.2125	0.0263	2.1900	2.2400	−1.609	0.0281	−1.633	−1.579
45	4	0.0018	0.149	−0.112	0.221	2.1875	0.0723	2.1200	2.2900	−1.2113	0.0688	−1.274	−1.113
50	4	−0.008	0.444	−0.376	0.517	2.095	0.250	1.920	2.450	−0.969	0.243	−1.139	−0.623
55	4	0.034	0.355	−0.292	0.341	1.8850	0.1912	1.7000	2.0500	−0.8925	0.1835	−1.068	−0.734
60	4	0.0948	0.1237	−0.012	0.214	1.6975	0.0709	1.6500	1.8000	−0.8455	0.0723	−0.894	−0.741
65	4	0.1008	0.0638	0.011	0.162	1.7575	0.00957	1.7500	1.7700	−0.5907	0.00776	−0.597	−0.581
70	4	−0.117	0.269	−0.35	0.118	1.3075	0.1715	1.1600	1.4800	−0.865	0.1672	−1.009	−0.7
75	4	−0.1048	0.0544	−0.154	−0.027	1.3525	0.0695	1.2700	1.4400	−0.6725	0.067	−0.756	−0.592
80	4	0.0577	0.0494	0.018	0.12	1.3425	0.1011	1.2500	1.4300	−0.562	0.0994	−0.653	−0.476
85	4	0.0885	0.0437	0.023	0.111	1.3150	0.1109	1.1500	1.3900	−0.4695	0.1099	−0.633	−0.395
90	4	−0.0892	0.1354	−0.204	0.107	1.0450	0.1634	0.8500	1.2500	−0.6375	0.1601	−0.83	−0.438
95	4	−0.0938	0.1177	−0.181	0.08	1.0100	0.1010	0.9500	1.1600	−0.5837	0.1005	−0.645	−0.435
100	4	−0.024	0.1478	−0.188	0.101	0.8400	0.0346	0.8100	0.8700	−0.6697	0.032	−0.698	−0.642

RB(standard deviation)

		RB(sd,1)				RB(sd,2)				RB(sd,3)			
10	4	-1.385	0.242	-1.737	-1.226	4.363	0.249	3.999	4.523	-3.543	0.253	-3.914	-3.383
15	4	-1.61	0.388	-1.943	-1.208	2.313	0.357	2.006	2.676	-2.854	0.327	-3.135	-2.52
20	4	-0.7665	0.1477	-0.941	-0.644	2.043	0.1239	1.894	2.145	-1.8158	0.123	-1.962	-1.714
25	4	-0.8992	0.0773	-1.015	-0.857	1.444	0.0494	1.37	1.47	-1.6175	0.0463	-1.687	-1.594
30	4	-0.6282	0.1082	-0.696	-0.469	1.2975	0.0539	1.245	1.373	-1.252	0.0515	-1.306	-1.182
35	4	-0.3832	0.0743	-0.463	-0.32	1.2633	0.0619	1.197	1.316	-0.9203	0.0588	-0.982	-0.87
40	4	-0.6442	0.0912	-0.721	-0.541	0.7885	0.0361	0.74	0.816	-1.1103	0.0346	-1.157	-1.084
45	4	-0.2348	0.0552	-0.268	-0.153	0.947	0.01869	0.937	0.975	-0.743	0.01934	-0.753	-0.714
50	4	-0.512	0.256	-0.733	-0.275	0.7107	0.1405	0.594	0.876	-0.809	0.1396	-0.925	-0.645
55	4	-0.3438	0.1368	-0.495	-0.228	0.7148	0.066	0.635	0.769	-0.6658	0.0652	-0.744	-0.612
60	4	-0.3482	0.1106	-0.444	-0.249	0.5983	0.0671	0.543	0.679	-0.6658	0.0665	-0.72	-0.584
65	4	-0.3682	0.1089	-0.446	-0.215	0.6183	0.0698	0.561	0.703	-0.5473	0.0682	-0.603	-0.464
70	4	-0.3515	0.1858	-0.512	-0.175	0.4728	0.1185	0.371	0.593	-0.6087	0.1156	-0.708	-0.491
75	4	-0.311	0.00337	-0.313	-0.306	0.5165	0.01601	0.493	0.529	-0.4922	0.01552	-0.515	-0.48
80	4	-0.1358	0.021	-0.152	-0.108	0.5707	0.0546	0.52	0.618	-0.377	0.0543	-0.427	-0.33
85	4	-0.1735	0.0317	-0.19	-0.126	0.5238	0.0435	0.465	0.57	-0.3657	0.0426	-0.423	-0.32
90	4	-0.2552	0.0457	-0.284	-0.187	0.3927	0.0667	0.315	0.478	-0.447	0.0655	-0.523	-0.363
95	4	-0.2597	0.0815	-0.311	-0.138	0.3748	0.0682	0.334	0.476	-0.4203	0.067	-0.461	-0.321
100	4	-0.176	0.0721	-0.264	-0.117	0.321	0.01361	0.304	0.332	-0.4325	0.01415	-0.45	-0.421

(Continued)

Table 3.21 (Continued) Descriptive statistics: RB(m,1), RB(m,2), RB(m,3), RB(s²,1), ..., RB(cv,3)

RB(coefficient of variation)

n	Freq	RB(cv,1)				RB(cv,2)				RB(cv,3)			
		Mean	StDev	Min	Max	Mean	StDev	Min	Max	Mean	StDev	Min	Max
10	4	-0.522	0.345	-0.905	-0.081	4.9545	0.1867	4.724	5.167	-3.309	0.0878	-3.396	-3.228
15	4	-0.728	0.507	-1.284	-0.071	2.853	0.46	2.317	3.4	-2.703	0.381	-3.189	-2.315
20	4	0.002	0.236	-0.267	0.215	2.5565	0.1452	2.395	2.741	-1.6	0.0907	-1.733	-1.529
25	4	-0.2773	0.1781	-0.401	-0.016	1.809	0.1076	1.736	1.967	-1.4675	0.0724	-1.529	-1.373
30	4	0.006	0.276	-0.285	0.38	1.6445	0.1262	1.559	1.832	-1.1178	0.0976	-1.222	-0.99
35	4	-0.1083	0.1861	-0.344	0.111	1.433	0.1174	1.273	1.53	-0.863	0.0829	-0.96	-0.769
40	4	-0.305	0.221	-0.555	-0.035	0.9722	0.0547	0.922	1.05	-1.0322	0.0347	-1.072	-0.992
45	4	0.0485	0.1885	-0.13	0.263	1.1175	0.0972	1.027	1.214	-0.676	0.0497	-0.721	-0.613
50	4	-0.3792	0.1939	-0.57	-0.125	0.7748	0.1193	0.604	0.858	-0.811	0.1283	-0.984	-0.688
55	4	0.034	0.1047	-0.097	0.129	0.9405	0.0928	0.838	1.04	-0.564	0.0672	-0.63	-0.497
60	4	-0.1373	0.1382	-0.278	0.039	0.7265	0.0877	0.625	0.836	-0.6108	0.0786	-0.686	-0.514
65	4	-0.163	0.218	-0.412	0.118	0.7223	0.1331	0.54	0.86	-0.5207	0.1115	-0.678	-0.415
70	4	-0.2017	0.1658	-0.36	-0.019	0.5685	0.1066	0.462	0.669	-0.5722	0.1107	-0.669	-0.469
75	4	-0.1745	0.0704	-0.235	-0.073	0.5998	0.0336	0.57	0.646	-0.4665	0.0241	-0.501	-0.448
80	4	0.099	0.0951	-0.039	0.175	0.6767	0.0558	0.615	0.741	-0.3403	0.0537	-0.389	-0.287
85	4	-0.002	0.1217	-0.136	0.15	0.6348	0.0612	0.544	0.674	-0.322	0.0487	-0.389	-0.286
90	4	-0.1487	0.0457	-0.199	-0.095	0.4645	0.0703	0.383	0.554	-0.4195	0.0711	-0.499	-0.328
95	4	-0.1192	0.0982	-0.227	-0.01	0.4383	0.0601	0.37	0.499	-0.402	0.0491	-0.449	-0.334
100	4	-0.0478	0.1214	-0.181	0.111	0.3898	0.0909	0.326	0.524	-0.4108	0.0525	-0.452	-0.334

Table 3.22 Descriptive statistics: RB(m,1), RB(m,2), RB(m,3), RB(s²,1), ...), RB(cv,3) (Results for $\alpha = 2.5$)

β	Freq	RB(·,1)				RB(·,2)				RB(·,3)			
		Mean	StDev	Min	Max	Mean	StDev	Min	Max	Mean	StDev	Min	Max
RB(mean)													
		RB(m,1)				**RB(m,2)**				**RB(m,3)**			
2.5	19	−0.0103	0.0779	−0.167	0.12	−0.0068	0.0687	−0.136	0.106	−0.00458	0.042	−0.084	0.061
4.5	19	0.0244	0.1056	−0.135	0.234	0.0238	0.0865	−0.103	0.199	0.0124	0.0524	−0.067	0.106
RB(variance)													
		RB(s²,1)				**RB(s²,2)**				**RB(s²,3)**			
2.5	19	0.0325	0.1607	−0.254	0.268	2.851	2.622	0.810	11.340	−1.434	1.205	−5.181	−0.435
4.5	19	−0.0197	0.2631	−0.557	0.486	2.873	2.560	0.870	10.980	−1.404	1.247	−5.373	−0.425
RB(standard deviation)													
		RB(sd,1)				**RB(sd,2)**				**RB(sd,3)**			
2.5	19	−0.4654	0.4249	−1.737	−0.126	1.078	0.931	0.315	3.999	−1.027	0.907	−3.914	−0.321
4.5	19	−0.546	0.442	−1.943	−0.117	1.056	1	0.332	4.523	−1.046	0.874	−3.383	−0.33
RB(coefficient of variation)													
		RB(cv,1)				**RB(cv,2)**				**RB(cv,3)**			
2.5	19	0.0281	0.1586	−0.199	0.38	1.378	1.223	0.365	5.167	−0.908	0.787	−3.396	−0.327
4.5	19	−0.2127	0.3396	−1.284	0.183	1.234	1.117	0.452	5.019	−1.012	0.861	−3.239	−0.301

Table 3.23 Descriptive statistics: RB(m,1), RB(m,2), RB(m,3), RB(s²,1), ...), RB(cv,3) (Results for α = 4.5)

RB(mean)

β	Freq	Mean	StDev	Min	Max
RB(m,1)					
2.5	19	-0.0164	0.0622	-0.149	0.07
4.5	19	-0.0025	0.0738	-0.128	0.15
RB(m,2)					
2.5	19	-0.0146	0.0513	-0.127	0.053
4.5	19	-0.0074	0.056	-0.108	0.085
RB(m,3)					
2.5	19	-0.00753	0.03084	-0.067	0.033
4.5	19	-0.00511	0.03403	-0.065	0.049

RB(variance)

β	Freq	Mean	StDev	Min	Max
RB(s²,1)					
2.5	19	-0.0197	0.2631	-0.557	0.486
4.5	19	-0.0002	0.1899	-0.292	0.517
RB(s²,2)					
2.5	19	2.873	2.560	0.870	10.980
4.5	19	2.866	2.529	0.810	10.940
RB(s²,3)					
2.5	19	-1.404	1.247	-5.373	-0.425
4.5	19	-1.411	1.272	-5.407	-0.395

RB(standard deviation)

β	Freq	Mean	StDev	Min	Max
RB(sd,1)					
2.5	19	-0.546	0.442	-1.943	-0.117
4.5	19	-0.5023	0.3866	-1.35	-0.108
RB(sd,2)					
2.5	19	1.056	1	0.332	4.523
4.5	19	1.078	0.985	0.304	4.408
RB(sd,3)					
2.5	19	-1.046	0.874	-3.383	-0.33
4.5	19	-1.026	0.859	-3.49	-0.32

RB(cv)

β	Freq	Mean	StDev	Min	Max
RB(cv,1)					
2.5	19	-0.2413	0.2596	-0.887	0.111
4.5	19	-0.2319	0.2363	-0.905	0.144
RB(cv,2)					
2.5	19	1.252	1.117	0.344	4.908
4.5	19	1.247	1.09	0.326	4.724
RB(cv,3)					
2.5	19	-0.954	0.813	-3.228	-0.286
4.5	19	-0.96	0.816	-3.373	-0.286

Scatterplot of RB(m,1), RB(m,2), RB(m,3), RB(v,1), RB(v,2), ... vs n

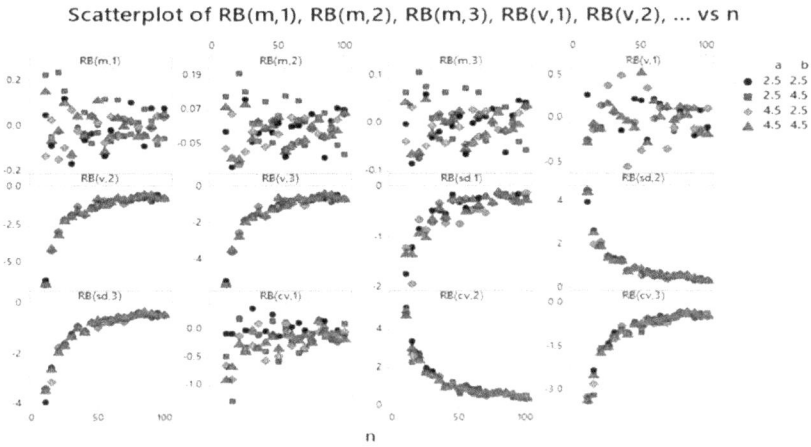

Figure 3.4 Scatter plots of RB vs sample size for mean (m), variance (s^2), standard deviation (sd) and coefficient of variation (cv).

Boxplot of RB(m,1), RB(m,2), RB(m,3), RB(v,1), RB(v,2), RB(v,3), ...

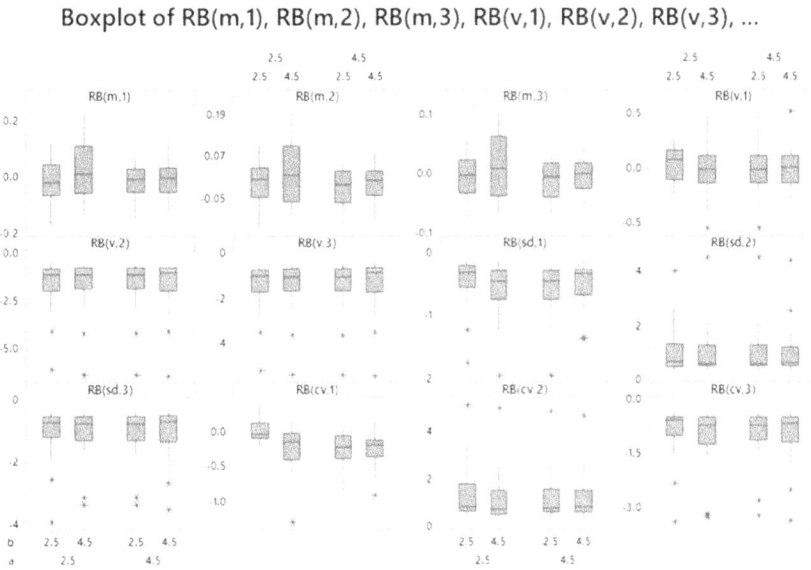

Figure 3.5 Box plots of RB vs different pairs of α and β for mean (m), variance (s^2), standard deviation (sd) and coefficient of variation (cv).

In Figure 3.4, we provide a graphical representation of relative bias values versus sample sizes for the four parameters mean, variance, standard deviation and coefficient of variation for four combinations of α and β.

In Figure 3.5, we provide box plots of relative bias values versus different pairs of α and β for the four parameters mean, variance, standard deviation and coefficient of variation for four combinations of α and β.

3.8 CONCLUSION

From the simulation study, we conclude that in the case of reasonably large bootstrapping samples, both SB and ISB perform better than CB not only for estimating mean but also for other parameters such as variance, standard deviation and coefficient of variation.

3.9 REMARKS

It is important to note that in the ISB sampling method, the units in the sample do not occur more than once; they are unique units of the sample population or not as they are replaced with the value of the sample mean of the values contained in the sample population. This is different when compared to the CB or SB sampling method in which the value can reoccur more than once, provided they are unique units of the sample population. The original version of this chapter can be found in Olalekan [17].

ACKNOWLEDGEMENTS

The authors are thankful to the Editor Dr. Sinha and referee for very constructive comments on the original version of this chapter. The authors are also thankful to Mr. Oluwatosin Lawal, SACM program student, for rechecking the original version of the manuscript.

REFERENCES

1. Singh, S. and Sedory, S. (2011). Sufficient bootstrapping. Computational Statistics and Data Analysis, 1629–1637.
2. Effron, B. (1992). Jackknife-after-bootstrap standards errors and influence functions. Journal of the Royal Statistical Society, 0964–1982.
3. Beyaztas, U. and Alin, A. (2013). Sufficient Jackknife after bootstrap sampling method. Communication Stat Papers, 1256–1267.
4. Beyaztas, U. and Alin, A. (2014). Sufficient Jackknife- after-bootstrap method for detection of influential observations in linear regression models. Stat Papers, 1001–1018.
5. Alin, A., Martin, M. A., Beyaztas, U. and Pathak, P. K. (2017). Sufficient M -out-of-n (m/n) bootstrap. Journal of Statistical Computation and Simulation, 1742–1753.
6. Beyaztas, U., Aylin, A. and Bandyopadhyay, S. (2022). Iterated sufficient M-out-of-N (M/n) bootstrap for non regular smooth function models. Journal of Data Science, 16(3), 593–604.
7. Dérian, N., Bellier, B., Pham, H.P, Tsitoura, E., Kazazi, D., Huret, C., Mavromara, P., Klatzmann, D. and Six, A. (2016). Early transcriptome

signatures from immunized mouse dendritic cells predict late vaccine-induced T-cell responses. PLoS Computational Biology, 1–17.

8. Beyaztas, H. B. (2016). An empirical comparison of block bootstrap method: Traditional and newer ones. Journal of Data Science, 641–656.

9. Beyaztas, U. and Beste, H. B. (2019). On Jackknife-after-bootstrap method for dependent data. Computational Econ, 1613–1632.

10. Beyaztas, U., Alin, A. and Soutir, B. (2018). Iterated sufficient M -out-of-N (M/N) bootstrap for non regular smooth function models. Journal of Data Science, 593–604.

11. Beyaztas, U. (2018). Assessment of individual bioequivalence using sufficient bootstrap procedure. Hartlin Turkey: Wiley.

12. Mostafa, S. A. and Ahmed, I. A. (2021). Kernel density estimation based on the distinct units in sampling with replacement. Sankhya, 83, 507–547.

13. Feller, W. (1957). An introduction to probability theory and application. NY: John Wiley and Sons.

14. Basu, D. (1957). On sampling with and without replacement. Sankhya, 1933–1960.

15. Raj, D. and Khamis, S. H. (1958). Some remarks on sampling with replacement. Annals of Mathematical Statistics, 550–557.

16. Pathak, P. K. (1961). On the evaluation of moments of distinct units in a sample. Sankhya, A, 23, 415–420.

17. Olalekan, H. (2020). An improved sufficient bootstrapping. Unpublished SACM project submitted to the Department of Mathematics, Texas A&M University-Kingsville, Kingsville, TX.

Chapter 4

A new measure of empirical mode

Steven Chavez, Stephen A. Sedory, and Sarjinder Singh

4.1 INTRODUCTION

Let y_i, $i = 1, 2, \cdots, N$ be the value of the *ith* unit in a population Ω. Let $\bar{Y} = \frac{1}{N} \Sigma_{i=1}^{N} y_i$ be the population mean of the study variable, y. Let $M_y = Q_{2y}$ be the median (second quartile) of the study variable, y. Doodson [1] was the first to show that, for a moderately skewed distribution, the empirical relationship between the median and mean is given by

$$Empirical\ Mode = 3M_y - 2\bar{Y} \qquad (4.1)$$

Sedory and Singh [2] developed ratio and regression-type estimators for estimating the empirical mode in the presence of an auxiliary variable. We know that sometimes the population mean is not known, however, it is more convenient to know the values of the three quartiles. Also we note that sometimes for highly positively skewed distributions, the empirical mode in (4.1) takes a negative value for a non-negative variable like having a gamma distribution. In general, $Q_{1y} + Q_{3y} < 2\bar{Y}$, so it helps to develop an empirical relationship between Q_{1y}, Q_{2y}, and Q_{3y} to find a new empirical mode and reduce the chances of taking a negative value.

We define a new measure of empirical mode as

$$M_{0y} = Mode_y = 3Q_{2y} - \left(Q_{1y} + Q_{3y}\right) \qquad (4.2)$$

Note that if

$$\bar{Y} = \frac{\left(Q_{1y} + Q_{3y}\right)}{2}, \qquad (4.3)$$

then the relation $M_{0y} = Empirical\ Mode$ holds.

DOI: 10.1201/9781003356653-4

In Appendix A, we provide some standard notations and expected values used in the rest of the chapter to study the properties of the three new estimators viz. naïve estimator, ratio estimator and regression type estimator in the following three sections. Sections 4.2, 4.3 and 4.4 are devoted to a naïve estimator, a ratio type estimator and a regression type estimator, respectively, and Section 4.5 concludes with a simulation study.

4.2 NAIVE ESTIMATOR

We propose a naive estimator of the newly proposed empirical mode M_{0y} as

$$\hat{M}_{0y} = 3\hat{Q}_{2y} - \hat{Q}_{1y} - \hat{Q}_{3y} \tag{4.4}$$

Then we have the following theorems.

Theorem 4.1: The proposed naive estimator $\hat{M}_{0y}$ is an unbiased estimator of M_{0y}.

Proof: Using Appendix A1, the estimator $\hat{M}_{0y}$ is terms of δ_{1y}, δ_{2y} and δ_{3y} can be written as

$$\hat{M}_{0y} = 3Q_{2y}(1 + \delta_{2y}) - Q_{1y}(1 + \delta_{1y}) - Q_{3y}(1 + \delta_{3y}) \tag{4.5}$$

Taking expected values on both sides of (4.5), we have

$$
\begin{aligned}
E[\hat{M}_{0y}] &= E[3Q_{2y}(1 + \delta_{2y}) - Q_{1y}(1 + \delta_{1y}) - Q_{3y}(1 + \delta_{3y})] \\
&= 3Q_{2y}[1 + E(\delta_{2y})] - Q_{1y}[1 + E(\delta_{1y})] - Q_{3y}[1 + E(\delta_{3y})] \\
&= 3Q_{2y} - Q_{1y} - Q_{3y} = M_{0y}
\end{aligned}
$$

so $\hat{M}_{0y}$ is an unbiased estimator of M_{0y}.

Theorem 4.2: The variance of the naïve estimator $\hat{M}_{0y}$ is given by

$$V(\hat{M}_{0y}) = \left(\frac{1-f}{n}\right)\left[\frac{9}{4\{f_y(Q_{2y})\}^2} + \frac{3}{16\{f_y(Q_{1y})\}^2} + \frac{3}{16\{f_y(Q_{3y})\}^2} - \frac{6}{8f_y(Q_{1y})f_y(Q_{2y})} \right.$$
$$\left. - \frac{6}{8f_y(Q_{2y})f_y(Q_{3y})} + \frac{1}{8f_y(Q_{1y})f_y(Q_{3y})}\right] \tag{4.6}$$

Proof: By the definition of variance, we have

$$
\begin{aligned}
V(\hat{M}_{0y}) &= E[\hat{M}_{0y} - E(\hat{M}_{0y})]^2 \\
&= E[3Q_{2y}(1 + \delta_{2y}) - Q_{1y}(1 + \delta_{1y}) - Q_{3y}(1 + \delta_{3y}) - M_{0y}]^2 \\
&= E[3Q_{2y} + 3Q_{2y}\delta_{2y} - Q_{1y} - Q_{1y}\delta_{1y} - Q_{3y} - Q_{3y}\delta_{3y} - M_{0y}]^2 \\
&= E[(3Q_{2y} - Q_{1y} - Q_{3y}) + 3Q_{2y}\delta_{2y} - Q_{1y}\delta_{1y} - Q_{3y}\delta_{3y} - M_{0y}]^2 \\
&= E[M_{0y} + 3Q_{2y}\delta_{2y} - Q_{1y}\delta_{1y} - Q_{3y}\delta_{3y} - M_{0y}]^2 \\
&= E[3Q_{2y}\delta_{2y} - Q_{1y}\delta_{1y} - Q_{3y}\delta_{3y}]^2 \\
&= E\Big[9Q_{2y}^2\delta_{2y}^2 + Q_{1y}^2\delta_{1y}^2 + Q_{3y}^2\delta_{3y}^2 - 6Q_{1y}Q_{2y}\delta_{1y}\delta_{2y} \\
&\quad - 6Q_{2y}Q_{3y}\delta_{2y}\delta_{3y} + 2Q_{1y}Q_{3y}\delta_{1y}\delta_{3y}\Big] \\
&= 9Q_{2y}^2 E\big(\delta_{2y}^2\big) + Q_{1y}^2 E\big(\delta_{1y}^2\big) + Q_{3y}^2 E\big(\delta_{3y}^2\big) - 6Q_{1y}Q_{2y}E(\delta_{1y}\delta_{2y}) \\
&\quad - 6Q_{2y}Q_{3y}E(\delta_{2y}\delta_{3y}) + 2Q_{1y}Q_{3y}E(\delta_{1y}\delta_{3y})\Big] \\
&= \left(\frac{1-f}{n}\right)\left(\frac{9}{4}\right)Q_{2y}^2\frac{1}{Q_{2y}^2\{f_y(Q_{2y})\}^2} + \left(\frac{1-f}{n}\right)\left(\frac{3}{16}\right)Q_{1y}^2\frac{1}{Q_{1y}^2\{f_y(Q_{1y})\}^2} \\
&\quad + \left(\frac{1-f}{n}\right)\left(\frac{3Q_{3y}^2}{16}\right)\frac{1}{Q_{3y}^2\{f_y(Q_{3y})\}^2} - 6Q_{1y}Q_{2y}\left(\frac{1-f}{8n}\right)\frac{1}{f_y(Q_{1y})f_y(Q_{2y})Q_{1y}Q_{2y}} \\
&\quad - 6Q_{2y}Q_{3y}\left(\frac{1-f}{8n}\right)\frac{1}{f_y(Q_{2y})f_y(Q_{3y})Q_{2y}Q_{3y}} \\
&\quad + 2Q_{1y}Q_{3y}\left(\frac{1-f}{16n}\right)\frac{1}{f_y(Q_{1y})f_y(Q_{3y})Q_{1y}Q_{3y}} \\
&= \left(\frac{1-f}{n}\right)\Big[\frac{9}{4\{f_y(Q_{2y})\}^2} + \frac{3}{16\{f_y(Q_{1y})\}^2} + \frac{3}{16\{f_y(Q_{3y})\}^2} - \frac{6}{8f_y(Q_{1y})f_y(Q_{2y})} \\
&\quad - \frac{6}{8f_y(Q_{2y})f_y(Q_{3y})} + \frac{2}{16f_y(Q_{1y})f_y(Q_{3y})}\Big] \\
&= \left(\frac{1-f}{n}\right)\Big[\frac{9}{4\{f_y(Q_{2y})\}^2} + \frac{3}{16\{f_y(Q_{1y})\}^2} + \frac{3}{16\{f_y(Q_{3y})\}^2} - \frac{3}{4f_y(Q_{1y})f_y(Q_{2y})} \\
&\quad - \frac{3}{4f_y(Q_{2y})f_y(Q_{3y})} + \frac{1}{8f_y(Q_{1y})f_y(Q_{3y})}\Big]
\end{aligned}
$$

which proves the theorem.

4.3 RATIO TYPE ESTIMATOR

Let $M_{0x} = 3Q_{2x} - Q_{1x} - Q_{3x}$ be the known value of the newly defined empirical mode for the auxiliary variable, x_i. Let $\hat{M}_{0x} = 3\hat{Q}_{2x} - \hat{Q}_{1x} - \hat{Q}_{3x}$ be the naive estimator of the newly defined empirical mode M_{0x}. Following Cochran [3], we define a ratio type estimator of the new measure of empirical mode as

$$\hat{M}_{0Ratio} = \hat{M}_{0y}\frac{M_{0x}}{\hat{M}_{0x}} \tag{4.7}$$

Now we have the following theorem.

Theorem 4.3: The bias in the ratio type estimator $\hat{M}_{0Ratio}$, to the first order of approximation, is given by

$$
\begin{aligned}
B[\hat{M}_{0Ratio}] &= \frac{M_{0y}}{M_{0x}^2}\left(\frac{1-f}{n}\right)\left[\frac{9}{4\{f_x(Q_{2x})\}^2} + \frac{3}{16\{f_x(Q_{1x})\}^2}\right.\\
&\quad + \frac{3}{16\{f_x(Q_{3x})\}^2} - \frac{3}{4f_x(Q_{1x})f_x(Q_{2x})}\\
&\quad \left. - \frac{3}{4f_x(Q_{2x})f_x(Q_{3x})} + \frac{1}{8f_x(Q_{1x})f_x(Q_{3x})}\right]\\
&\quad - \frac{1}{M_{0x}}\left(\frac{1-f}{n}\right)\left[\frac{9\left(P_{Q_{2x}Q_{2y}}-\frac{1}{4}\right)}{f_x(Q_{2x})f_y(Q_{2y})} - \frac{3\left(P_{Q_{2x}Q_{1y}}-\frac{1}{8}\right)}{f_x(Q_{2x})f_y(Q_{1y})}\right.\\
&\quad - \frac{3\left(P_{Q_{2x}Q_{3y}}-\frac{3}{8}\right)}{f_x(Q_{2x})f_y(Q_{3y})} - \frac{3\left(P_{Q_{1x}Q_{2y}}-\frac{1}{8}\right)}{f_x(Q_{1x})f_y(Q_{2y})}\\
&\quad + \frac{\left(P_{Q_{1x}Q_{1y}}-\frac{1}{16}\right)}{f_x(Q_{1x})f_y(Q_{1y})} + \frac{\left(P_{Q_{1x}Q_{3y}}-\frac{3}{16}\right)}{f_x(Q_{1x})f_y(Q_{3y})} - \frac{3\left(P_{Q_{3x}Q_{2y}}-\frac{3}{8}\right)}{f_x(Q_{3x})f_y(Q_{2y})}\\
&\quad \left. + \frac{\left(P_{Q_{3x}Q_{1y}}-\frac{3}{16}\right)}{f_x(Q_{3x})f_y(Q_{1y})} + \frac{\left(P_{Q_{3x}Q_{3y}}-\frac{9}{16}\right)}{f_x(Q_{3x})f_y(Q_{3y})}\right]
\end{aligned} \tag{4.8}
$$

Proof: Using Appendix A1, the ratio type estimator $\hat{M}_{0Ratio}$ can be approximated as

$$
\begin{aligned}
\hat{M}_{0Ratio} &= \hat{M}_{0y}\frac{M_{0x}}{\hat{M}_{0x}}\\
&= (3\hat{Q}_{2y} - \hat{Q}_{1y} - \hat{Q}_{3y})\frac{M_{0x}}{(3\hat{Q}_{2x} - \hat{Q}_{1x} - \hat{Q}_{3x})}\\
&= [3Q_{2y}(1 + \delta_{2y}) - Q_{1y}(1 + \delta_{1y}) - Q_{3y}(1 + \delta_{3y})]\\
&\quad \times \frac{M_{0x}}{3Q_{2x}(1 + \delta_{2x}) - Q_{1x}(1 + \delta_{1x}) - Q_{3x}(1 + \delta_{3x})}\\
&= [(3Q_{2y} - Q_{1y} - Q_{3y}) + 3Q_{2y}\delta_{2y} - Q_{1y}\delta_{1y} - Q_{3y}\delta_{3y}]\\
&\quad \times \frac{M_{0x}}{(3Q_{2x} - Q_{1x} - Q_{3x}) + 3Q_{2x}\delta_{2x} - Q_{1x}\delta_{1x} - Q_{3x}\delta_{3x}}
\end{aligned}
$$

$$= [M_{0y} + 3Q_{2y}\delta_{2y} - Q_{1y}\delta_{1y} - Q_{3y}\delta_{3y}]\frac{M_{0x}}{M_{0x} + 3Q_{2x}\delta_{2x} - Q_{1x}\delta_{1x} - Q_{3x}\delta_{3x}}$$

$$= [M_{0y} + 3Q_{2y}\delta_{2y} - Q_{1y}\delta_{1y} - Q_{3y}\delta_{3y}]\frac{1}{\left[1 + \frac{3Q_{2x}\delta_{2x} - Q_{1x}\delta_{1x} - Q_{3x}\delta_{3x}}{M_{0x}}\right]}$$

$$= M_{0y}\left[1 + \frac{3Q_{2y}\delta_{2y} - Q_{1y}\delta_{1y} - Q_{3y}\delta_{3y}}{M_{0y}}\right]\left[1 + \frac{3Q_{2x}\delta_{2x} - Q_{1x}\delta_{1x} - Q_{3x}\delta_{3x}}{M_{0x}}\right]^{-1}$$

$$= M_{0y}\left[1 + \frac{3Q_{2y}\delta_{2y} - Q_{1y}\delta_{1y} - Q_{3y}\delta_{3y}}{M_{0y}}\right]\left[1 - \frac{3Q_{2x}\delta_{2x} - Q_{1x}\delta_{1x} - Q_{3x}\delta_{3x}}{M_{0x}}\right.$$

$$\left. + \left(\frac{3Q_{2x}\delta_{2x} - Q_{1x}\delta_{1x} - Q_{3x}\delta_{3x}}{M_{0x}}\right)^2 + O_p(n^{-1})\right]$$

$$= M_{0y}\left[1 + \frac{3Q_{2y}\delta_{2y} - Q_{1y}\delta_{1y} - Q_{3y}\delta_{3y}}{M_{0y}} - \frac{3Q_{2x}\delta_{2x} - Q_{1x}\delta_{1x} - Q_{3x}\delta_{3x}}{M_{0x}}\right.$$

$$+ \left(\frac{3Q_{2x}\delta_{2x} - Q_{1x}\delta_{1x} - Q_{3x}\delta_{3x}}{M_{0x}}\right)^2$$

$$\left. - \frac{(3Q_{2y}\delta_{2y} - Q_{1y}\delta_{1y} - Q_{3y}\delta_{3y})(3Q_{2x}\delta_{2x} - Q_{1x}\delta_{1x} - Q_{3x}\delta_{3x})}{M_{0y}M_{0x}} + O_p(n^{-1})\right]$$

$$= M_{0y}\left[1 + \frac{3Q_{2y}\delta_{2y} - Q_{1y}\delta_{1y} - Q_{3y}\delta_{3y}}{M_{0y}} - \frac{3Q_{2x}\delta_{2x} - Q_{1x}\delta_{1x} - Q_{3x}\delta_{3x}}{M_{0x}}\right.$$

$$+ \frac{9Q_{2x}^2\delta_{2x}^2 + Q_{1x}^2\delta_{1x}^2 + Q_{3x}^2\delta_{3x}^2 - 6Q_{1x}Q_{2x}\delta_{1x}\delta_{2x} - 6Q_{2x}Q_{3x}\delta_{2x}\delta_{3x} + 2Q_{1x}Q_{3x}\delta_{1x}\delta_{3x}}{M_{0x}^2}$$

$$- \left\{\frac{9Q_{2y}Q_{2x}\delta_{2y}\delta_{2x} - 3Q_{1y}Q_{2x}\delta_{1y}\delta_{2x} - 3Q_{2x}Q_{3y}\delta_{2x}\delta_{3y} - 3Q_{1x}Q_{2y}\delta_{1x}\delta_{2y}}{M_{0y}M_{0x}}\right.$$

$$\left. + \frac{Q_{1x}Q_{1y}\delta_{1x}\delta_{1y} + Q_{1x}Q_{3y}\delta_{1x}\delta_{3y} - 3Q_{2y}Q_{3x}\delta_{2y}\delta_{3x} + Q_{1y}Q_{3x}\delta_{1y}\delta_{3x} + Q_{3x}Q_{3y}\delta_{3x}\delta_{3y}}{M_{0y}M_{0x}}\right\}$$

$$\left. + O_p(n^{-1})\right]$$

which holds true under the assumption $\left|\dfrac{3Q_{2x}\delta_{2x} - Q_{1x}\delta_{1x} - Q_{3x}\delta_{3x}}{M_{0x}}\right| < 1.$
Taking expected values on both sides, we get

$$E[\hat{M}_{0Ratio}] = M_{0y}\left[1 + 0 - 0 + \frac{1}{M_{0x}^2}\left\{9Q_{2x}^2 E(\delta_{2x}^2) + Q_{1x}^2 E(\delta_{1x}^2) + Q_{3x}^2 E(\delta_{3x}^2)\right.\right.$$

$$\left. - 6Q_{1x}Q_{2x}E(\delta_{1x}\delta_{2x}) - 6Q_{2x}Q_{3x}E(\delta_{2x}\delta_{3x}) + 2Q_{1x}Q_{3x}E(\delta_{1x}\delta_{3x})\right\}$$

$$- \frac{1}{M_{0y}M_{0x}}\left\{9Q_{2y}Q_{2x}E(\delta_{2y}\delta_{2x}) - 3Q_{1y}Q_{2x}E(\delta_{1y}\delta_{2x})\right.$$

$$- 3Q_{2x}Q_{3y}E(\delta_{2x}\delta_{3y}) - 3Q_{1x}Q_{2y}E(\delta_{1x}\delta_{2y}) + Q_{1x}Q_{1y}E(\delta_{1x}\delta_{1y})$$

$$+ Q_{1x}Q_{3y}E(\delta_{1x}\delta_{3y})$$

$$- 3Q_{2y}Q_{3x}E\left(\delta_{2y}\delta_{3x}\right) + Q_{1y}Q_{3x}E\left(\delta_{1y}\delta_{3x}\right) + Q_{3x}Q_{3y}E\left(\delta_{3x}\delta_{3y}\right)\big\}$$
$$+ O_p\left(n^{-1}\right)\Big]$$

$$= M_{0y} + \frac{M_{0y}}{M_{0x}^2}\left[9Q_{2x}^2\left(\frac{1-f}{n}\right)\frac{1}{4Q_{2x}^2\{f_x(Q_{2x})\}^2} + Q_{1x}^2\left(\frac{1-f}{n}\right)\frac{3}{16Q_{1x}^2\{f_x(Q_{1x})\}^2}\right.$$

$$+ Q_{3x}^2\left(\frac{1-f}{n}\right)\frac{3}{16Q_{3x}^2\{f_x(Q_{3x})\}^2} - 6Q_{1x}Q_{2x}\left(\frac{1-f}{8n}\right)\frac{1}{f_x(Q_{1x})f_x(Q_{2x})Q_{1x}Q_{2x}}$$

$$\left. - 6Q_{2x}Q_{3x}\left(\frac{1-f}{8n}\right)\frac{1}{f_x(Q_{2x})f_x(Q_{3x})Q_{2x}Q_{3x}} + 2Q_{1x}Q_{3x}\left(\frac{1-f}{16n}\right)\frac{1}{f_x(Q_{1x})f_x(Q_{3x})Q_{1x}Q_{3x}}\right]$$

$$- \frac{1}{M_{0x}}\left[9Q_{2y}Q_{2x}\left(\frac{1-f}{n}\right)\frac{\{f_x(Q_{2x})\}^{-1}\{f_y(Q_{2y})\}^{-1}}{Q_{2x}Q_{2y}}\left(P_{Q_{2x}Q_{2y}} - \frac{1}{4}\right)\right.$$

$$- 3Q_{1y}Q_{2x}\left(\frac{1-f}{n}\right)\frac{\{f_x(Q_{2x})\}^{-1}\{f_y(Q_{1y})\}^{-1}}{Q_{2x}Q_{1y}}\left(P_{Q_{2x}Q_{1y}} - \frac{1}{8}\right)$$

$$- 3Q_{2x}Q_{3y}\left(\frac{1-f}{n}\right)\frac{\{f_x(Q_{2x})\}^{-1}\{f_y(Q_{3y})\}^{-1}}{Q_{2x}Q_{3y}}\left(P_{Q_{2x}Q_{3y}} - \frac{3}{8}\right)$$

$$- 3Q_{1x}Q_{2y}\left(\frac{1-f}{n}\right)\frac{\{f_x(Q_{1x})\}^{-1}\{f_y(Q_{2y})\}^{-1}}{Q_{1x}Q_{2y}}\left(P_{Q_{1x}Q_{2y}} - \frac{1}{8}\right)$$

$$+ Q_{1x}Q_{1y}\left(\frac{1-f}{n}\right)\frac{\{f_x(Q_{1x})\}^{-1}\{f_y(Q_{1y})\}^{-1}}{Q_{1x}Q_{1y}}\left(P_{Q_{1x}Q_{1y}} - \frac{1}{16}\right)$$

$$+ Q_{1x}Q_{3y}\left(\frac{1-f}{n}\right)\frac{\{f_x(Q_{1x})\}^{-1}\{f_y(Q_{3y})\}^{-1}}{Q_{1x}Q_{3y}}\left(P_{Q_{1x}Q_{3y}} - \frac{3}{16}\right)$$

$$- 3Q_{2y}Q_{3x}\left(\frac{1-f}{n}\right)\frac{\{f_x(Q_{3x})\}^{-1}\{f_y(Q_{2y})\}^{-1}}{Q_{3x}Q_{2y}}\left(P_{Q_{3x}Q_{2y}} - \frac{3}{8}\right)$$

$$+ Q_{1y}Q_{3x}\left(\frac{1-f}{n}\right)\frac{\{f_x(Q_{3x})\}^{-1}\{f_y(Q_{1y})\}^{-1}}{Q_{1y}Q_{3x}}\left(P_{Q_{3x}Q_{1y}} - \frac{3}{16}\right)$$

$$\left. + Q_{3x}Q_{3y}\left(\frac{1-f}{n}\right)\frac{\{f_x(Q_{3x})\}^{-1}\{f_y(Q_{3y})\}^{-1}}{Q_{3x}Q_{3y}}\left(P_{Q_{3x}Q_{3y}} - \frac{9}{16}\right) + O\left(n^{-1}\right)\right]$$

$$= M_{0y} + \frac{M_{0y}}{M_{0x}^2}\left(\frac{1-f}{n}\right)\left[\frac{9}{4\{f_x(Q_{2x})\}^2} + \frac{3}{16\{f_x(Q_{1x})\}^2} + \frac{3}{16\{f_x(Q_{3x})\}^2} - \frac{3}{4f_x(Q_{1x})f_x(Q_{2x})}\right.$$

$$\left. - \frac{3}{4f_x(Q_{2x})f_x(Q_{3x})} + \frac{1}{8f_x(Q_{1x})f_x(Q_{3x})}\right] - \frac{1}{M_{0x}}\left(\frac{1-f}{n}\right)\left[\frac{9\left(P_{Q_{2x}Q_{2y}} - \frac{1}{4}\right)}{f_x(Q_{2x})f_y(Q_{2y})}\right.$$

$$- \frac{3\left(P_{Q_{2x}Q_{1y}} - \frac{1}{8}\right)}{f_x(Q_{2x})f_y(Q_{1y})} - \frac{3\left(P_{Q_{2x}Q_{3y}} - \frac{3}{8}\right)}{f_x(Q_{2x})f_y(Q_{3y})} - \frac{3\left(P_{Q_{1x}Q_{2y}} - \frac{1}{8}\right)}{f_x(Q_{1x})f_y(Q_{2y})}$$

$$+ \frac{\left(P_{Q_{1x}Q_{1y}} - \frac{1}{16}\right)}{f_x(Q_{1x})f_y(Q_{1y})} + \frac{\left(P_{Q_{1x}Q_{3y}} - \frac{3}{16}\right)}{f_x(Q_{1x})f_y(Q_{3y})} - \frac{3\left(P_{Q_{3x}Q_{2y}} - \frac{3}{8}\right)}{f_x(Q_{3x})f_y(Q_{2y})}$$

$$\left. + \frac{\left(P_{Q_{3x}Q_{1y}} - \frac{3}{16}\right)}{f_x(Q_{3x})f_y(Q_{1y})} + \frac{\left(P_{Q_{3x}Q_{3y}} - \frac{9}{16}\right)}{f_x(Q_{3x})f_y(Q_{3y})}\right]$$

which proves the theorem.

Clearly $B[\hat{M}_{0Ratio}] \to 0$ as $n \to N$. Thus, $\hat{M}_{0Ratio}$ is a consistent estimator of M_{0y}.

Theorem 4.4: The mean squared error of the ratio type estimator $\hat{M}_{0Ratio}$, to the first order of approximation, is given by

$$
\begin{aligned}
MSE[\hat{M}_{0Ratio}] = & \left(\frac{1-f}{n}\right)\Bigg[\Bigg\{\frac{9}{4\{f_y(Q_{2y})\}^2} + \frac{3}{16\{f_y(Q_{1y})\}^2} + \frac{3}{16\{f_y(Q_{3y})\}^2} - \frac{3}{4f_y(Q_{1y})f_y(Q_{2y})} \\
& - \frac{3}{4f_y(Q_{2y})f_y(Q_{3y})} + \frac{1}{8f_y(Q_{1y})f_y(Q_{3y})}\Bigg\} \\
& + \left(\frac{M_{0y}}{M_{0x}}\right)^2 \Bigg\{\frac{9}{4\{f_x(Q_{2x})\}^2} + \frac{3}{16\{f_x(Q_{1x})\}^2} + \frac{3}{16\{f_x(Q_{3x})\}^2} \\
& - \frac{3}{4f_x(Q_{1x})f_x(Q_{2x})} - \frac{3}{4f_x(Q_{2x})f_x(Q_{3x})} + \frac{1}{8f_x(Q_{1x})f_x(Q_{3x})}\Bigg\} \\
& - 2\left(\frac{M_{0y}}{M_{0x}}\right)\Bigg\{\frac{9\left(P_{Q_{2x}Q_{2y}}-\frac{1}{4}\right)}{f_x(Q_{2x})f_y(Q_{2y})} - \frac{3\left(P_{Q_{2x}Q_{1y}}-\frac{1}{8}\right)}{f_x(Q_{2x})f_y(Q_{1y})} - \frac{3\left(P_{Q_{2x}Q_{3y}}-\frac{3}{8}\right)}{f_x(Q_{2x})f_y(Q_{3y})} \\
& - \frac{3\left(P_{Q_{1x}Q_{2y}}-\frac{1}{8}\right)}{f_x(Q_{1x})f_y(Q_{2y})} + \frac{\left(P_{Q_{1x}Q_{1y}}-\frac{1}{16}\right)}{f_x(Q_{1x})f_y(Q_{1y})} \\
& + \frac{\left(P_{Q_{1x}Q_{3y}}-\frac{3}{16}\right)}{f_x(Q_{1x})f_y(Q_{3y})} - \frac{3\left(P_{Q_{3x}Q_{2y}}-\frac{3}{8}\right)}{f_x(Q_{3x})f_y(Q_{2y})} + \frac{\left(P_{Q_{3x}Q_{1y}}-\frac{3}{16}\right)}{f_x(Q_{3x})f_y(Q_{1y})} \\
& + \frac{\left(P_{Q_{3x}Q_{3y}}-\frac{9}{16}\right)}{f_x(Q_{3x})f_y(Q_{3y})}\Bigg\}\Bigg]
\end{aligned}
$$

(4.9)

Proof: The mean squared error of the proposed ratio type estimator is given by

$$
\begin{aligned}
MSE[\hat{M}_{0Ratio}] &= E\left[\hat{M}_{0Ratio} - M_{0y}\right]^2 \\
&= E\left[M_{0y}\left\{1 + \frac{3Q_{2y}\delta_{2y} - Q_{1y}\delta_{1y} - Q_{3y}\delta_{3y}}{M_{0y}}\right.\right. \\
&\qquad - \frac{3Q_{2x}\delta_{2x} - Q_{1x}\delta_{1x} - Q_{3x}\delta_{3x}}{M_{0x}} \\
&\qquad \left.\left.+ O_p(n^{-1})\right\} - M_{0y}\right]^2 \\
&= M_{0y}^2 E\left[\frac{3Q_{2y}\delta_{2y} - Q_{1y}\delta_{1y} - Q_{3y}\delta_{3y}}{M_{0y}}\right. \\
&\qquad - \frac{3Q_{2x}\delta_{2x} - Q_{1x}\delta_{1x} - Q_{3x}\delta_{3x}}{M_{0x}} \\
&\qquad \left.+ O_p(n^{-1})\right]^2 \\
&= M_{0y}^2 E\left[\frac{(3Q_{2y}\delta_{2y} - Q_{1y}\delta_{1y} - Q_{3y}\delta_{3y})^2}{M_{0y}^2} + \frac{(3Q_{2x}\delta_{2x} - Q_{1x}\delta_{1x} - Q_{3x}\delta_{3x})^2}{M_{0x}^2}\right.
\end{aligned}
$$

$$- \frac{2\left(3Q_{2y}\delta_{2y} - Q_{1y}\delta_{1y} - Q_{3y}\delta_{3y}\right)\left(3Q_{2x}\delta_{2x} - Q_{1x}\delta_{1x} - Q_{3x}\delta_{3x}\right)}{M_{0y}M_{0x}} + O_p\left(n^{-1}\right) \Bigg]$$

$$= M_{0y}^2 E\Bigg[\frac{9Q_{2y}^2\delta_{2y}^2 + Q_{1y}^2\delta_{1y}^2 + Q_{3y}^2\delta_{3y}^2 - 6Q_{1y}Q_{2y}\delta_{1y}\delta_{2y} - 6Q_{2y}Q_{3y}\delta_{2y}\delta_{3y} + 2Q_{1y}Q_{3y}\delta_{1y}\delta_{3y}}{M_{0y}^2}$$

$$+ \frac{9Q_{2x}^2\delta_{2x}^2 + Q_{1x}^2\delta_{1x}^2 + Q_{3x}^2\delta_{3x}^2 - 6Q_{1x}Q_{2x}\delta_{1x}\delta_{2x} - 6Q_{2x}Q_{3x}\delta_{2x}\delta_{3x} + 2Q_{1x}Q_{3x}\delta_{1x}\delta_{3x}}{M_{0x}^2}$$

$$- \frac{2\{9Q_{2x}Q_{2y}\delta_{2x}\delta_{2y} - 3Q_{1y}Q_{2x}\delta_{2x}\delta_{1y} - 3Q_{2x}Q_{3y}\delta_{2x}\delta_{3y} - 3Q_{1x}Q_{2y}\delta_{1x}\delta_{2y} + Q_{1x}Q_{1y}\delta_{1x}\delta_{1y}}{M_{0x}M_{0y}}$$

$$+ \frac{Q_{1x}Q_{3y}\delta_{1x}\delta_{3y} - 3Q_{3x}Q_{2y}\delta_{3x}\delta_{2y} + Q_{3x}Q_{1y}\delta_{3x}\delta_{1y} + Q_{3x}Q_{3y}\delta_{3x}\delta_{3y}\}}{M_{0x}M_{0y}} \Bigg\} + O_p\left(n^{-1}\right) \Bigg]$$

$$= 9Q_{2y}^2 E\left(\delta_{2y}^2\right) + Q_{1y}^2 E\left(\delta_{1y}^2\right) + Q_{3y}^2\left(\delta_{3y}^2\right) - 6Q_{1y}Q_{2y}E\left(\delta_{1y}\delta_{2y}\right)$$

$$- 6Q_{2y}Q_{3y}E\left(\delta_{2y}\delta_{3y}\right) + 2Q_{1y}Q_{3y}E\left(\delta_{1y}\delta_{3y}\right)$$

$$+ \left(\frac{M_{0y}}{M_{0x}}\right)^2 \Big[9Q_{2x}^2 E\left(\delta_{2x}^2\right) + Q_{1x}^2 E\left(\delta_{1x}^2\right) + Q_{3x}^2\left(\delta_{3x}^2\right) - 6Q_{1x}Q_{2x}E\left(\delta_{1x}\delta_{2x}\right)$$

$$- 6Q_{2x}Q_{3x}E\left(\delta_{2x}\delta_{3x}\right) + 2Q_{1x}Q_{3x}E\left(\delta_{1x}\delta_{3x}\right) \Big]$$

$$- 2\left(\frac{M_{0y}}{M_{0x}}\right)\Big[9Q_{2x}Q_{2y}E\left(\delta_{2x}\delta_{2y}\right) - 3Q_{1y}Q_{2x}E\left(\delta_{2x}\delta_{1y}\right) - 3Q_{2x}Q_{3y}E\left(\delta_{2x}\delta_{3y}\right)$$

$$- 3Q_{1x}Q_{2y}E\left(\delta_{1x}\delta_{2y}\right) + Q_{1x}Q_{1y}E\left(\delta_{1x}\delta_{1y}\right) + Q_{1x}Q_{3y}E\left(\delta_{1x}\delta_{3y}\right)$$

$$- 3Q_{3x}Q_{2y}E\left(\delta_{3x}\delta_{2y}\right) + Q_{3x}Q_{1y}E\left(\delta_{3x}\delta_{1y}\right) + Q_{3x}Q_{3y}E\left(\delta_{3x}\delta_{3y}\right) + O\left(n^{-1}\right) \Big]$$

$$= \Bigg[9Q_{2y}^2\left(\frac{1-f}{4n}\right)\frac{1}{Q_{2y}^2\{f_y(Q_{2y})\}^2} + Q_{1y}^2\left(\frac{1-f}{n}\right)\frac{3}{16Q_{1y}^2\{f_y(Q_{1y})\}^2} + Q_{3y}^2\left(\frac{1-f}{n}\right)\frac{3}{16Q_{3y}^2\{f_y(Q_{3y})\}^2}$$

$$- 6Q_{1y}Q_{2y}\left(\frac{1-f}{8n}\right)\frac{1}{f_y(Q_{1y})f_y(Q_{2y})Q_{1y}Q_{2y}} - 6Q_{2y}Q_{3y}\left(\frac{1-f}{8n}\right)\frac{1}{f_y(Q_{2y})f_y(Q_{3y})Q_{2y}Q_{3y}}$$

$$+ 2Q_{1y}Q_{3y}\left(\frac{1-f}{16n}\right)\frac{1}{f_y(Q_{1y})f_y(Q_{3y})Q_{1y}Q_{3y}} \Bigg] + \left(\frac{M_{0y}}{M_{0x}}\right)^2\Bigg[9Q_{2x}^2\left(\frac{1-f}{4n}\right)\frac{1}{Q_{2x}^2\{f_x(Q_{2x})\}^2}$$

$$+ Q_{1x}^2\left(\frac{1-f}{n}\right)\left(\frac{3}{16}\right)\frac{1}{Q_{1x}^2\{f_x(Q_{1x})\}^2} + Q_{3x}^2\left(\frac{1-f}{n}\right)\left(\frac{3}{16}\right)\frac{1}{Q_{3x}^2\{f_x(Q_{3x})\}^2}$$

$$- 6Q_{1x}Q_{2x}\left(\frac{1-f}{8n}\right)\frac{1}{f_x(Q_{1x})f_x(Q_{2x})Q_{1x}Q_{2x}} - 6Q_{2x}Q_{3x}\left(\frac{1-f}{8n}\right)\frac{1}{f_x(Q_{2x})f_x(Q_{3x})Q_{2x}Q_{3x}}$$

$$+ 2Q_{1x}Q_{3x}\left(\frac{1-f}{16n}\right)\frac{1}{f_x(Q_{1x})f_x(Q_{3x})Q_{1x}Q_{3x}} \Bigg]$$

$$- 2\left(\frac{M_{0y}}{M_{0x}}\right)\Bigg[9Q_{2y}Q_{2x}\left(\frac{1-f}{n}\right)\frac{\{f_x(Q_{2x})\}^{-1}\{f_y(Q_{2y})\}^{-1}}{Q_{2x}Q_{2y}}\left(P_{Q_{2x}Q_{2y}} - \frac{1}{4}\right)$$

$$- 3Q_{1y}Q_{2x}\left(\frac{1-f}{n}\right)\frac{\{f_x(Q_{2x})\}^{-1}\{f_y(Q_{1y})\}^{-1}}{Q_{2x}Q_{1y}}\left(P_{Q_{2x}Q_{1y}} - \frac{1}{8}\right)$$

$$- 3Q_{2x}Q_{3y}\left(\frac{1-f}{n}\right)\frac{\{f_x(Q_{2x})\}^{-1}\{f_y(Q_{3y})\}^{-1}}{Q_{2x}Q_{3y}}\left(P_{Q_{2x}Q_{3y}} - \frac{3}{8}\right)$$

$$
- 3Q_{1x}Q_{2y}\left(\frac{1-f}{n}\right)\frac{\{f_x(Q_{1x})\}^{-1}\{f_y(Q_{2y})\}^{-1}}{Q_{1x}Q_{2y}}\left(P_{Q_{1x}Q_{2y}} - \frac{1}{8}\right)
$$

$$
+ Q_{1x}Q_{1y}\left(\frac{1-f}{n}\right)\frac{\{f_x(Q_{1x})\}^{-1}\{f_y(Q_{1y})\}^{-1}}{Q_{1x}Q_{1y}}\left(P_{Q_{1x}Q_{1y}} - \frac{1}{16}\right)
$$

$$
+ Q_{1x}Q_{3y}\left(\frac{1-f}{n}\right)\frac{\{f_x(Q_{1x})\}^{-1}\{f_y(Q_{3y})\}^{-1}}{Q_{1x}Q_{3y}}\left(P_{Q_{1x}Q_{3y}} - \frac{3}{16}\right)
$$

$$
- 3Q_{2y}Q_{3x}\left(\frac{1-f}{n}\right)\frac{\{f_x(Q_{3x})\}^{-1}\{f_y(Q_{2y})\}^{-1}}{Q_{3x}Q_{2y}}\left(P_{Q_{3x}Q_{2y}} - \frac{3}{8}\right)
$$

$$
+ Q_{1y}Q_{3x}\left(\frac{1-f}{n}\right)\frac{\{f_x(Q_{3x})\}^{-1}\{f_y(Q_{1y})\}^{-1}}{Q_{1y}Q_{3x}}\left(P_{Q_{3x}Q_{1y}} - \frac{3}{16}\right)
$$

$$
\left.+ Q_{3x}Q_{3y}\left(\frac{1-f}{n}\right)\frac{\{f_x(Q_{3x})\}^{-1}\{f_y(Q_{3y})\}^{-1}}{Q_{3x}Q_{3y}}\left(P_{Q_{3x}Q_{3y}} - \frac{9}{16}\right)\right]
$$

$$
= \left(\frac{1-f}{n}\right)\left[\left\{\frac{9}{4\{f_y(Q_{2y})\}^2} + \frac{3}{16\{f_y(Q_{1y})\}^2} + \frac{3}{16\{f_y(Q_{3y})\}^2}\right.\right.
$$

$$
- \frac{3}{4f_y(Q_{1y})f_y(Q_{2y})} - \frac{3}{4f_y(Q_{2y})f_y(Q_{3y})}
$$

$$
+ \frac{1}{8f_y(Q_{1y})f_y(Q_{3y})}\right\} + \left(\frac{M_{0y}}{M_{0x}}\right)^2\left\{\frac{9}{4\{f_x(Q_{2x})\}^2} + \frac{3}{16\{f_x(Q_{1x})\}^2}\right.
$$

$$
+ \frac{3}{16\{f_x(Q_{3x})\}^2} - \frac{3}{4f_x(Q_{1x})f_x(Q_{2x})}
$$

$$
- \frac{3}{4f_x(Q_{2x})f_x(Q_{3x})} + \frac{1}{8f_x(Q_{1x})f_x(Q_{3x})}\right\}
$$

$$
- 2\left(\frac{M_{0y}}{M_{0x}}\right)\left[\frac{9\left(P_{Q_{2x}Q_{2y}} - \frac{1}{4}\right)}{f_x(Q_{2x})f_y(Q_{2y})} - \frac{3\left(P_{Q_{2x}Q_{1y}} - \frac{1}{8}\right)}{f_x(Q_{2x})f_y(Q_{1y})}\right.
$$

$$
- \frac{3\left(P_{Q_{2x}Q_{3y}} - \frac{3}{8}\right)}{f_x(Q_{2x})f_y(Q_{3y})} - \frac{3\left(P_{Q_{1x}Q_{2y}} - \frac{1}{8}\right)}{f_x(Q_{1x})f_y(Q_{2y})}
$$

$$
+ \frac{\left(P_{Q_{1x}Q_{1y}} - \frac{1}{16}\right)}{f_x(Q_{1x})f_y(Q_{1y})} + \frac{\left(P_{Q_{1x}Q_{3y}} - \frac{3}{16}\right)}{f_x(Q_{1x})f_y(Q_{3y})}
$$

$$
\left.\left.\left.- \frac{3\left(P_{Q_{3x}Q_{2y}} - \frac{3}{8}\right)}{f_x(Q_{3x})f_y(Q_{2y})} + \frac{\left(P_{Q_{3x}Q_{1y}} - \frac{3}{16}\right)}{f_x(Q_{3x})f_y(Q_{1y})} + \frac{\left(P_{Q_{3x}Q_{3y}} - \frac{9}{16}\right)}{f_x(Q_{3x})f_y(Q_{3y})}\right]\right]\right]
$$

which proves the theorem.

4.4 REGRESSION TYPE ESTIMATOR

Following Hansen, Hurwitz and Madow [4], we define a regression type estimator of the new measure of empirical mode as

$$
\hat{M}_{0Reg} = \hat{M}_{0y} + \beta\left(M_{0x} - \hat{M}_{0x}\right) \tag{4.10}
$$

where β is a constant to be determined such that the variance of estimator is minimum. We have the following theorems.

Theorem 4.5: For any given β, the estimator $\hat{M}_{0Reg}$ is an unbiased estimator of the new measure of the empirical mode, M_{0y}.

Proof: Using Appendix A1, the estimator $\hat{M}_{0Reg}$ can be written as

$$
\begin{aligned}
\hat{M}_{0Reg} &= \hat{M}_{0y} + \beta[M_{0x} - \hat{M}_{0x}] \\
&= \left(3\hat{Q}_{2y} - \hat{Q}_{1y} - \hat{Q}_{3y}\right) + \beta\left[M_{0x} - \left(3\hat{Q}_{2x} - \hat{Q}_{1x} - \hat{Q}_{3x}\right)\right] \\
&= 3Q_{2y}(1 + \delta_{2y}) - Q_{1y}(1 + \delta_{1y}) - Q_{3y}(1 + \delta_{3y}) \\
&\quad + \beta[M_{0x} - \{3Q_{2x}(1 + \delta_{2x}) - Q_{1x}(1 + \delta_{1x}) - Q_{3x}(1 + \delta_{3x})\}] \\
&= (3Q_{2y} - Q_{1y} - Q_{3y}) + 3Q_{2y}\delta_{2y} - Q_{1y}\delta_{1y} - Q_{3y}\delta_{3y} \\
&\quad + \beta[M_{0x} - \{3Q_{2x} - Q_{1x} - Q_{3x} + 3Q_{2x}\delta_{2x} - Q_{1x}\delta_{1x} \\
&\quad - Q_{3x}\delta_{3x}\}] \\
&= M_{0y} + 3Q_{2y}\delta_{2y} - Q_{1y}\delta_{1y} - Q_{3y}\delta_{3y} \\
&\quad + \beta[M_{0x} - \{M_{0x} + 3Q_{2x}\delta_{2x} - Q_{1x}\delta_{1x} - Q_{3x}\delta_{3x}\}] \\
&= M_{0y} + \{3Q_{2y}\delta_{2y} - Q_{1y}\delta_{1y} - Q_{3y}\delta_{3y}\} \\
&\quad - \beta\{3Q_{2x}\delta_{2x} - Q_{1x}\delta_{1x} - Q_{3x}\delta_{3x}\}
\end{aligned}
\tag{4.11}
$$

taking the expected values on both sides of (4.11), we get

$$
\begin{aligned}
E[\hat{M}_{0Reg}] &= M_{0y} + E[\{3Q_{2y}\delta_{2y} - Q_{1y}\delta_{1y} - Q_{3y}\delta_{3y}\} \\
&\quad - \beta\{3Q_{2x}\delta_{2x} - Q_{1x}\delta_{1x} - Q_{3x}\delta_{3x}\}] \\
&= M_{0y} + 3Q_{2y}E(\delta_{2y}) - Q_{1y}E(\delta_{1y}) - Q_{3y}E(\delta_{3y}) \\
&\quad - \beta\{3Q_{2x}E(\delta_{2x}) - Q_{1x}E(\delta_{1x}) - Q_{3x}E(\delta_{3x})\} \\
&= M_{0y} + 0 + 0 = M_{0y}
\end{aligned}
$$

which proves the theorem.

Theorem 4.6: The minimum variance of the proposed regression type estimator $\hat{M}_{0Reg}$ is given by

$$
Min.\ V(\hat{M}_{0Reg}) = \left(\frac{1-f}{n}\right)\left[\left\{\frac{9}{4\{f_y(Q_{2y})\}^2} + \frac{3}{16\{f_y(Q_{1y})\}^2}\right.\right.
$$

$$
+ \frac{3}{16\{f_y(Q_{3y})\}^2} - \frac{3}{4f_y(Q_{1y})f_y(Q_{2y})}
$$

$$
- \frac{3}{4f_y(Q_{2y})f_y(Q_{3y})} + \frac{1}{8f_y(Q_{1y})f_y(Q_{3y})}\right\}
$$

$$
- \left\{\frac{9\left(P_{Q_{2x}Q_{2y}}-\frac{1}{4}\right)}{f_x(Q_{2x})f_y(Q_{2y})} - \frac{3\left(P_{Q_{2x}Q_{1y}}-\frac{1}{8}\right)}{f_x(Q_{2x})f_y(Q_{1y})} - \frac{3\left(P_{Q_{2x}Q_{3y}}-\frac{3}{8}\right)}{f_x(Q_{2x})f_y(Q_{3y})}\right.
$$

$$
- \frac{3\left(P_{Q_{1x}Q_{2y}}-\frac{1}{8}\right)}{f_x(Q_{1x})f_y(Q_{2y})} + \frac{\left(P_{Q_{1x}Q_{1y}}-\frac{1}{16}\right)}{f_x(Q_{1x})f_y(Q_{1y})}
$$

$$
+ \frac{\left(P_{Q_{1x}Q_{3y}}-\frac{3}{16}\right)}{f_x(Q_{1x})f_y(Q_{3y})} - \frac{3\left(P_{Q_{3x}Q_{2y}}-\frac{3}{8}\right)}{f_x(Q_{3x})f_y(Q_{2y})} + \frac{\left(P_{Q_{3x}Q_{1y}}-\frac{3}{16}\right)}{f_x(Q_{3x})f_y(Q_{1y})}
$$

$$
+ \frac{\left(P_{Q_{3x}Q_{3y}}-\frac{9}{16}\right)}{f_x(Q_{3x})f_y(Q_{3y})}\}^2 \div \left\{\frac{9}{4\{f_x(Q_{2x})\}^2} + \frac{3}{16\{f_x(Q_{1x})\}^2}\right.
$$

$$
+ \frac{3}{16\{f_x(Q_{3x})\}^2} - \frac{3}{4f_x(Q_{1x})f_x(Q_{2x})}
$$

$$
\left.\left.- \frac{3}{4f_x(Q_{2x})f_x(Q_{3x})} + \frac{1}{8f_x(Q_{1x})f_x(Q_{3x})}\right\}\right]
$$

$$(4.12)$$

Proof: The variance of the proposed estimator $\hat{M}_{0Reg}$ is given by

$$
V(\hat{M}_{0Reg}) = E[\hat{M}_{0Reg} - E(\hat{M}_{0Reg})]^2
$$

$$
= E[(3Q_{2y}\delta_{2y} - Q_{1y}\delta_{1y} - Q_{3y}\delta_{3y}) - \beta(3Q_{2x}\delta_{2x} - Q_{1x}\delta_{1x} - Q_{3x}\delta_{3x})]^2
$$

$$
= E[(3Q_{2y}\delta_{2y} - Q_{1y}\delta_{1y} - Q_{3y}\delta_{3y})^2 + \beta^2(3Q_{2x}\delta_{2x} - Q_{1x}\delta_{1x} - Q_{3x}\delta_{3x})^2
$$

$$
- 2\beta(3Q_{2y}\delta_{2y} - Q_{1y}\delta_{1y} - Q_{3y}\delta_{3y})(3Q_{2x}\delta_{2x} - Q_{1x}\delta_{1x} - Q_{3x}\delta_{3x})]
$$

$$
= E\left[9Q_{2y}^2\delta_{2y}^2 + Q_{1y}^2\delta_{1y}^2 + Q_{3y}^2\delta_{3y}^2 - 6Q_{1y}Q_{2y}\delta_{1y}\delta_{2y} - 6Q_{2y}Q_{3y}\delta_{2y}\delta_{3y}\right.
$$

$$
+ 2Q_{1y}Q_{3y}\delta_{1y}\delta_{3y} + \beta^2\left\{9Q_{2x}^2\delta_{2x}^2 + Q_{1x}^2\delta_{1x}^2 + Q_{3x}^2\delta_{3x}^2 - 6Q_{1x}Q_{2x}\delta_{1x}\delta_{2x}\right.
$$

$$
- 6Q_{2x}Q_{3x}\delta_{2x}\delta_{3x} + 2Q_{1x}Q_{3x}\delta_{1x}\delta_{3x}\} - 2\beta\{9Q_{2y}Q_{2x}\delta_{2y}\delta_{2x}
$$

$$
- 3Q_{1y}Q_{2x}\delta_{1y}\delta_{2x} - 3Q_{2x}Q_{3y}\delta_{2x}\delta_{3y} + Q_{1x}Q_{1y}\delta_{1x}\delta_{1y} - 3Q_{1x}Q_{2y}\delta_{1x}\delta_{2y}
$$

$$
\left.+ Q_{1x}Q_{3y}\delta_{1x}\delta_{3y} - 3Q_{2y}Q_{3x}\delta_{2y}\delta_{3x} + Q_{1y}Q_{3x}\delta_{1y}\delta_{3x} + Q_{3y}Q_{3x}\delta_{3y}\delta_{3x}\}\right]
$$

$$
\begin{aligned}
= \; & 9Q_{2y}^2 E(\delta_{2y}^2) + Q_{1y}^2 E(\delta_{1y}^2) + Q_{3y}^2 E(\delta_{3y}^2) - 6Q_{1y}Q_{2y}E(\delta_{1y}\delta_{2y}) \\
& - 6Q_{2y}Q_{3y}E(\delta_{2y}\delta_{3y}) + 2Q_{1y}Q_{3y}E(\delta_{1y}\delta_{3y}) + \beta^2\Big\{9Q_{2x}^2 E(\delta_{2x}^2) + Q_{1x}^2 E(\delta_{1x}^2) \\
& + Q_{3x}^2 E(\delta_{3x}^2) - 6Q_{1x}Q_{2x}E(\delta_{1x}\delta_{2x}) - 6Q_{2x}Q_{3x}E(\delta_{2x}\delta_{3x}) + 2Q_{1x}Q_{3x}E(\delta_{1x}\delta_{3x})\Big\} \\
& - 2\beta\Big\{9Q_{2y}Q_{2x}E(\delta_{2y}\delta_{2x}) - 3Q_{1y}Q_{2x}E(\delta_{1y}\delta_{2x}) - 3Q_{1x}Q_{2y}E(\delta_{1x}\delta_{2y}) \\
& - 3Q_{2x}Q_{3y}E(\delta_{2x}\delta_{3y}) + Q_{1x}Q_{1y}E(\delta_{1x}\delta_{1y}) + Q_{1x}Q_{3y}E(\delta_{1x}\delta_{3y}) \\
& - 3Q_{2y}Q_{3x}E(\delta_{2y}\delta_{3x}) + Q_{1y}Q_{3x}E(\delta_{1y}\delta_{3x}) + Q_{3y}Q_{3x}E(\delta_{3y}\delta_{3x})\}\Big] \\[6pt]
= \; & 9Q_{2y}^2\left(\frac{1-f}{n}\right)\frac{1}{4Q_{2y}^2\{f_y(Q_{2y})\}^2} + \left(\frac{1-f}{n}\right)\frac{3Q_{1y}^2}{16}\frac{1}{Q_{1y}^2\{f_y(Q_{1y})\}^2} + Q_{3y}^2\left(\frac{1-f}{n}\right)\frac{3}{16Q_{3y}^2\{f_y(Q_{3y})\}^2} \\
& - 6Q_{1y}Q_{2y}\left(\frac{1-f}{8n}\right)\frac{1}{f_y(Q_{1y})f_y(Q_{2y})Q_{1y}Q_{2y}} - 6Q_{2y}Q_{3y}\left(\frac{1-f}{8n}\right)\frac{1}{f_y(Q_{2y})f_y(Q_{3y})Q_{3y}Q_{3y}} \\
& + 2Q_{1y}Q_{3y}\left(\frac{1-f}{16n}\right)\frac{1}{f_y(Q_{1y})f_y(Q_{3y})Q_{1y}Q_{3y}} + \beta^2\Bigg[9Q_{2x}^2\left(\frac{1-f}{4n}\right)\frac{1}{Q_{2x}^2\{f_x(Q_{2x})\}^2} \\
& + Q_{1x}^2\left(\frac{1-f}{n}\right)\frac{3}{16Q_{1x}^2\{f_x(Q_{1x})\}^2} + Q_{3x}^2\left(\frac{1-f}{n}\right)\frac{3}{16Q_{3x}^2\{f_x(Q_{3x})\}^2} \\
& - 6Q_{1x}Q_{2x}\left(\frac{1-f}{8n}\right)\frac{1}{f_x(Q_{1x})f_x(Q_{2x})Q_{1x}Q_{2x}} - 6Q_{2x}Q_{3x}\left(\frac{1-f}{8n}\right)\frac{1}{f_x(Q_{2x})f_x(Q_{3x})Q_{2x}Q_{3x}} \\
& + 2Q_{1x}Q_{3x}\left(\frac{1-f}{16n}\right)\frac{1}{f_x(Q_{1x})f_x(Q_{3x})Q_{1x}Q_{3x}}\Bigg] \\
& - 2\beta\Bigg[9Q_{2y}Q_{2x}\left(\frac{1-f}{n}\right)\frac{\{f_x(Q_{2x})\}^{-1}\{f_y(Q_{2y})\}^{-1}}{Q_{2x}Q_{2y}}\left(P_{Q_{2x}Q_{2y}} - \frac{1}{4}\right) \\
& - 3Q_{1y}Q_{2x}\left(\frac{1-f}{n}\right)\frac{\{f_x(Q_{2x})\}^{-1}\{f_y(Q_{1y})\}^{-1}}{Q_{2x}Q_{1y}}\left(P_{Q_{2x}Q_{1y}} - \frac{1}{8}\right) \\
& - 3Q_{2x}Q_{3y}\left(\frac{1-f}{n}\right)\frac{\{f_x(Q_{2x})\}^{-1}\{f_y(Q_{3y})\}^{-1}}{Q_{2x}Q_{3y}}\left(P_{Q_{2x}Q_{3y}} - \frac{3}{8}\right) \\
& - 3Q_{1x}Q_{2y}\left(\frac{1-f}{n}\right)\frac{\{f_x(Q_{1x})\}^{-1}\{f_y(Q_{2y})\}^{-1}}{Q_{1x}Q_{2y}}\left(P_{Q_{1x}Q_{2y}} - \frac{1}{8}\right) \\
& + Q_{1x}Q_{1y}\left(\frac{1-f}{n}\right)\frac{\{f_x(Q_{1x})\}^{-1}\{f_y(Q_{1y})\}^{-1}}{Q_{1x}Q_{1y}}\left(P_{Q_{1x}Q_{1y}} - \frac{1}{16}\right) \\
& + Q_{1x}Q_{3y}\left(\frac{1-f}{n}\right)\frac{\{f_x(Q_{1x})\}^{-1}\{f_y(Q_{3y})\}^{-1}}{Q_{1x}Q_{3y}}\left(P_{Q_{1x}Q_{3y}} - \frac{3}{16}\right) \\
& - 3Q_{2y}Q_{3x}\left(\frac{1-f}{n}\right)\frac{\{f_x(Q_{3x})\}^{-1}\{f_y(Q_{2y})\}^{-1}}{Q_{3x}Q_{2y}}\left(P_{Q_{3x}Q_{2y}} - \frac{3}{8}\right) \\
& + Q_{1y}Q_{3x}\left(\frac{1-f}{n}\right)\frac{\{f_x(Q_{3x})\}^{-1}\{f_y(Q_{1y})\}^{-1}}{Q_{1y}Q_{3x}}\left(P_{Q_{3x}Q_{1y}} - \frac{3}{16}\right) \\
& + Q_{3x}Q_{3y}\left(\frac{1-f}{n}\right)\frac{\{f_x(Q_{3x})\}^{-1}\{f_y(Q_{3y})\}^{-1}}{Q_{3x}Q_{3y}}\left(P_{Q_{3x}Q_{3y}} - \frac{9}{16}\right)\Bigg] \\[6pt]
= \; & \left(\frac{1-f}{n}\right)\Bigg[\frac{9}{4\{f_y(Q_{2y})\}^2} + \frac{3}{16\{f_y(Q_{1y})\}^2} + \frac{3}{16\{f_y(Q_{3y})\}^2} - \frac{3}{4f_y(Q_{1y})f_y(Q_{2y})} \\
& - \frac{3}{4f_y(Q_{2y})f_y(Q_{3y})} + \frac{1}{8f_y(Q_{1y})f_y(Q_{3y})} + \beta^2\Bigg\{\frac{9}{4\{f_x(Q_{2x})\}^2} + \frac{3}{16\{f_x(Q_{1x})\}^2} \\
& + \frac{3}{16\{f_x(Q_{3x})\}^2} - \frac{3}{4f_x(Q_{1x})f_x(Q_{2x})} - \frac{3}{4f_x(Q_{2x})f_x(Q_{3x})} + \frac{1}{8f_x(Q_{1x})f_x(Q_{3x})}\Bigg\} \\
& - 2\beta\Bigg\{\frac{9\left(P_{Q_{2x}Q_{2y}} - \frac{1}{4}\right)}{f_x(Q_{2x})f_y(Q_{2y})} - \frac{3\left(P_{Q_{2x}Q_{1y}} - \frac{1}{8}\right)}{f_x(Q_{2x})f_y(Q_{1y})} - \frac{3\left(P_{Q_{2x}Q_{3y}} - \frac{3}{8}\right)}{f_x(Q_{2x})f_y(Q_{3y})} - \frac{3\left(P_{Q_{1x}Q_{2y}} - \frac{1}{8}\right)}{f_x(Q_{1x})f_y(Q_{2y})}
\end{aligned}
$$

$$+ \frac{\left(P_{Q_{1x}Q_{1y}} - \frac{1}{16}\right)}{f_x(Q_{1x})f_y(Q_{1y})} + \frac{\left(P_{Q_{1x}Q_{3y}} - \frac{3}{16}\right)}{f_x(Q_{1x})f_y(Q_{3y})} - \frac{3\left(P_{Q_{3x}Q_{2y}} - \frac{3}{8}\right)}{f_x(Q_{3x})f_y(Q_{2y})}$$

$$+ \frac{\left(P_{Q_{3x}Q_{1y}} - \frac{3}{16}\right)}{f_x(Q_{3x})f_y(Q_{1y})} + \frac{\left(P_{Q_{3x}Q_{3y}} - \frac{9}{16}\right)}{f_x(Q_{3x})f_y(Q_{3y})}\Bigg)\Bigg] \tag{4.13}$$

On differentiating $V\left[\hat{M}_{0Reg}\right]$ in (4.13) with respect to β, and equaling to zero, we get

$$\beta = \Bigg[\frac{9\left(P_{Q_{2x}Q_{2y}} - \frac{1}{4}\right)}{f_x(Q_{2x})f_y(Q_{2y})} - \frac{3\left(P_{Q_{2x}Q_{1y}} - \frac{1}{8}\right)}{f_x(Q_{2x})f_y(Q_{1y})} - \frac{3\left(P_{Q_{2x}Q_{3y}} - \frac{3}{8}\right)}{f_x(Q_{2x})f_y(Q_{3y})} - \frac{3\left(P_{Q_{1x}Q_{2y}} - \frac{1}{8}\right)}{f_x(Q_{1x})f_y(Q_{2y})}$$

$$+ \frac{\left(P_{Q_{1x}Q_{1y}} - \frac{1}{16}\right)}{f_x(Q_{1x})f_y(Q_{1y})} + \frac{\left(P_{Q_{1x}Q_{3y}} - \frac{3}{16}\right)}{f_x(Q_{1x})f_y(Q_{3y})} - \frac{3\left(P_{Q_{3x}Q_{2y}} - \frac{3}{8}\right)}{f_x(Q_{3x})f_y(Q_{2y})}$$

$$+ \frac{\left(P_{Q_{3x}Q_{1y}} - \frac{3}{16}\right)}{f_x(Q_{3x})f_y(Q_{1y})} + \frac{\left(P_{Q_{3x}Q_{3y}} - \frac{9}{16}\right)}{f_x(Q_{3x})f_y(Q_{3y})}\Bigg]$$

$$\div \Bigg[\frac{9}{4\{f_x(Q_{2x})\}^2} + \frac{3}{16\{f_x(Q_{1x})\}^2} + \frac{3}{16\{f_x(Q_{3x})\}^2} - \frac{3}{4f_x(Q_{1x})f_x(Q_{2x})}$$

$$- \frac{3}{4f_x(Q_{2x})f_x(Q_{3x})} + \frac{1}{8f_x(Q_{1x})f_x(Q_{3x})}\Bigg] \tag{4.14}$$

Now on substituting β from (4.14) into (4.13), we get the minimum $V\left(\hat{M}_{0Reg}\right)$ in (4.12), which proves the theorem.

In the next section, we provide empirical evidence by following the works of Eappen, Sedory and Singh [5,6] in the support of the proposed new measure of empirical mode and its three estimators.

4.5 EMPIRICAL EVIDENCE: SIMULATION

To study the performance of the newly proposed estimators we conducted a simulation study. Following Singh and Horn [7], we simulated several populations consisting of a study variable Y and auxiliary variable X each of size N=10,000 by using the transformation

$$y_i = \mu_y + \left(y_i^* - \mu_{y^*}\right)\sqrt{1 - \rho^2} + \frac{\sigma_y \rho \left(x_i^* - \mu_{x^*}\right)}{\sigma_x} \tag{4.15}$$

and the auxiliary variable,

$$x_i = \mu_x + \left(x_i^* - \mu_{x^*}\right) \tag{4.16}$$

where $y*$, and $x*$ are independent gamma variables. Assume a_y and b_y are the shape and scale parameters for the study variable, $y*$, and a_x and b_x are the shape and scale parameters for the variable, $x*$. In other words $y* \sim G(a_y, b_y)$ and $x* \sim G(a_x, b_x)$ to create a simulated population. In the Figure 4.1, we show the distribution of Y and X for each correlation value used from 0.80 to 0.98.

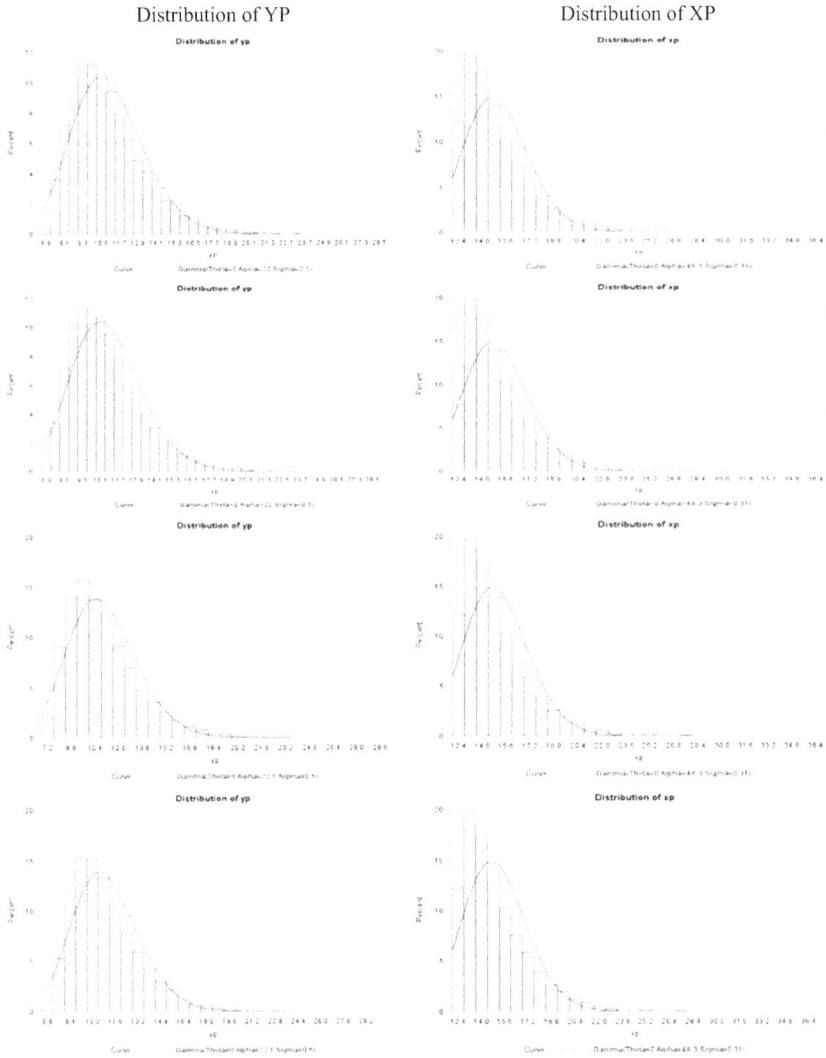

Figure 4.1 Distributions of y and x.

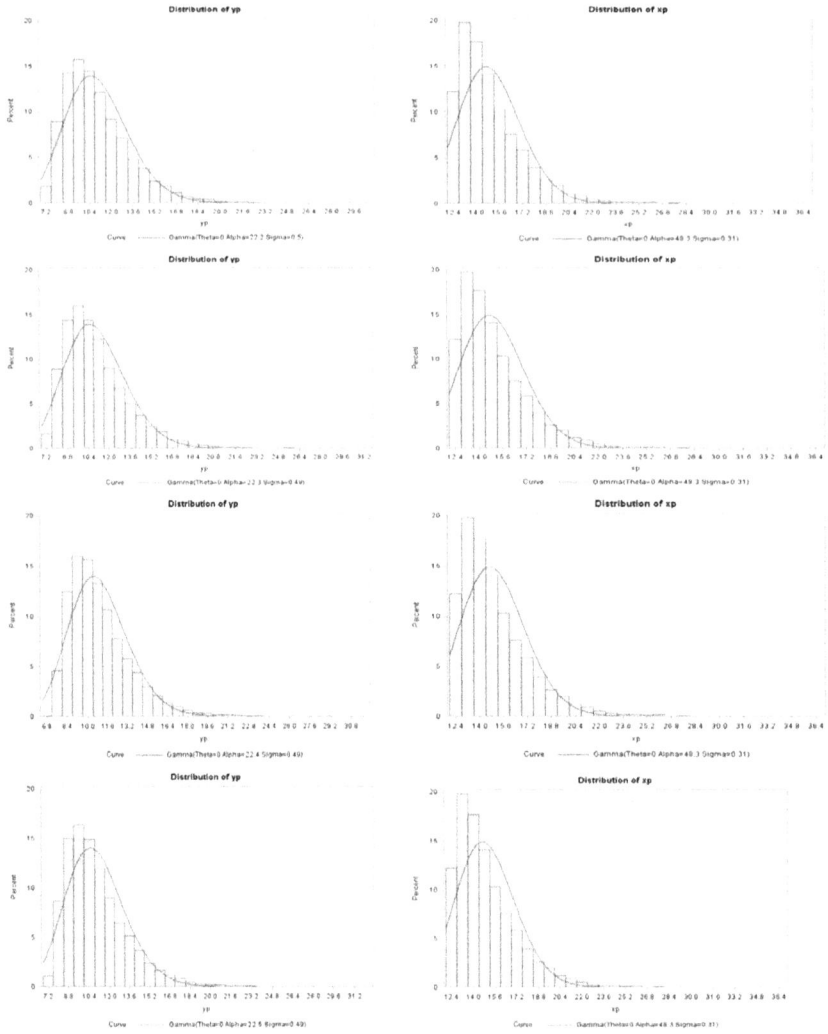

Figure 4.1   (continued)

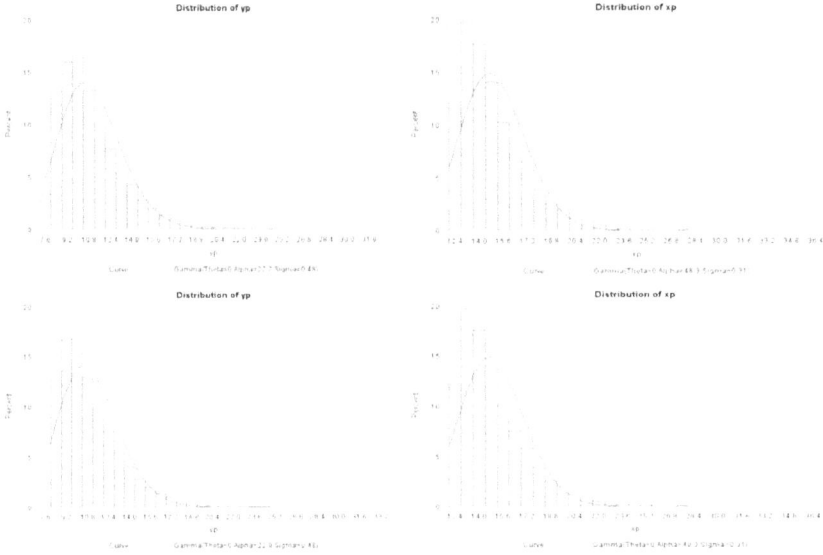

Figure 4.1   (continued)

After creating the population, we calculated the relative bias of the ratio estimator since there is no bias in the proposed naïve estimator or in the regression-type estimator. The relative bias in the ratio type estimator is computed as

$$RB(\hat{M}_{0Ratio}) = \frac{B[\hat{M}_{0Ratio}] \times 100\%}{M_{0y}} \tag{4.17}$$

We then calculated the relative efficiency of the ratio type estimator and the regression type estimator with respect to the naïve estimator. The relative efficiency for the ratio-type estimator with respect to the naïve estimator is given by

$$RE_1 = \frac{V(M_{0y}) \times 100\%}{MSE(\hat{M}_{0Ratio})} \tag{4.18}$$

The relative efficiency of the regression type with respect to the naïve estimator is given by

$$RE_2 = \frac{V(M_{0y}) \times 100\%}{Min.V(\hat{M}_{0Reg})} \tag{4.19}$$

The results from the simulation study can be seen in Table 4.1.

It may be worth pointing out that new fitted values $\hat{a}_y$, $\hat{b}_y$, $\hat{a}_x$ and $\hat{b}_x$ of a_y, b_y, a_x and b_x are used from the generated populations while computing the value of RB, RE_1 and RE_2 by following Eappen, Sedory and Singh [5,6]. From Table 4.2, one can observe that the fitted values of the shape and scale parameters for the variables y and x are quite different from those for y^* and x^*. Also there are variations in the values of $\hat{a}_y$, and $\hat{b}_y$ as the value of ρ varies from 0.80 to 0.98, but there is no change in the values of $\hat{a}_x$ and $\hat{b}_x$ as also reflected in the transformations used in (4.15) and (4.16).

The simulation study is performed using SAS codes given in Appendix A2. In Table 4.1 we have the input value of ρ, and observed value of ρ_{xy}, in each one of the ten created populations. The relative bias of the ratio type estimator and the relative efficiencies are calculated. The higher value of the

Table 4.1 Summary of RB and RE

ρ	ρ_{xy}	RB Ratio	RE_1 (%)	RE_2 (%)
0.80	0.80225	0.004003637	80.841	103.011
0.82	0.82200	0.003846384	82.683	103.801
0.84	0.84174	0.003687367	84.644	104.704
0.86	0.86149	0.003422041	88.123	106.475
0.88	0.88124	0.003194060	91.410	108.230
0.90	0.90099	0.002860575	96.673	111.202
0.92	0.92076	0.002466531	103.602	115.569
0.94	0.94054	0.001896177	115.476	123.885
0.96	0.96033	0.001268477	132.023	136.756
0.98	0.98014	0.000287774	169.302	169.686

Table 4.2 Fitted values of shape and scale parameters of the gamma distributions

ρ	$\hat{a}_y$	$\hat{b}_y$	$\hat{a}_x$	$\hat{b}_x$
0.80	21.9757	0.50013	48.266	0.31056
0.82	22.0137	0.49926	48.266	0.31056
0.84	22.0629	0.49813	48.266	0.31056
0.86	22.1245	0.49674	48.266	0.31056
0.88	22.2003	0.49503	48.266	0.31056
0.90	22.2922	0.49298	48.266	0.31056
0.92	22.4028	0.49054	48.266	0.31056
0.94	22.5357	0.48764	48.266	0.31056
0.96	22.6965	0.48418	48.266	0.31056
0.98	22.8954	0.47997	48.266	0.31056

correlation coefficient, ρ_{xy} leads to gain in more efficiency for the ratio and the regression estimators with respect to the naïve estimator. The relative bias of the ratio-type estimator is well below 10%. For the relative efficiency of the ratio type estimator, to get over 100% we needed a correlation of greater than 0.90. The relative efficiency of the regression estimator is over 100% on all cases of correlations above 0.80. The original version of this work is available in Chavez [8].

REFERENCES

1. Doodson, A.T. III. (1917) Relation of the mode, median and mean in frequency curves. Biometrika, *11*(4), 425–429.
2. Sedory, S.A. and Singh, S. (2014). Estimation of mode using auxiliary information. Communications in Statistics – Simulation and Computation, *43*(10), 2390–2402.
3. Cochran, W. G. (1940). The estimation of the yields of cereal experiment by sampling for the ratio of grain to total produce. The Journal of Agriculture Science, *30*(2), 262–275.
4. Hansen, M.H., Hurwitz, W. N. and Madow, W. G. (1953). Sample Survey Methods and Theory, Vol. 1. Wiley, New York.
5. Eappen, C.V., Sedory, S.A. and Singh, S. (2022a). Ratio and regression type estimators for a new measure of coefficient of dispersion relative to the empirical mode. Chapter 11 in Optimal Decision Making in Operations Research and Statistics: Methodologies and Applications. CRC Press. ISBN 9780367618759
6. Eappen, C.V., Sedory, S.A. and Singh, S. (2022b). Ratio and regression type estimators of a new measure of coefficient of dispersion. Communication in Statistics: Simulation and Computing, *51*(4), 1899–1920.
7. Singh, S. and Horn, S. (1998). An alternative estimator for multi-character surveys. Metrika, *48*(2), 99–107.
8. Chavez, S. (2020). Estimation of dispersion index using auxiliary information. Unpublished MS thesis submitted to the Department of Mathematics, Texas A&M University-Kingsville, Kingsville, TX.
9. Kuk, A.Y.C. and Mak, T. (1989). Median estimation in the presence of auxiliary information. Journal of the Royal Statistical Society, *51*(2), 261–269.
10. Singh, H.P., Solanki, R.S. and Singh, S. (2004). Estimation of Bowley's coefficient of skewness in the presence of auxiliary information, Communications in Statistics – Theory and Methods, *43*(22), 4867–4880.
11. Singh, H.P., Puertas, S.M. and Singh, S. (2006). Estimation of interquartile range of the study variable using the known interquartile range of auxiliary variable. International Journal of Applied Mathematics and Statistics, 6, 33–47.

APPENDIX A

A.I NOTATIONS AND EXPECTED VALUES

Let (y_i, x_i), $i = 1, 2, \ldots N$ be the ordered pairs of the values of the study variable, Y and the auxiliary variable, X in a population Ω. Again, let (y_i, x_i), $i = 1, 2 \ldots n$ be the ordered pairs of the values of the study variable, Y and the auxiliary variable, X in a simple random and without replacement sample, s_n, of size n which is abbreviated as SRSWOR. Let $y_{(1)} \leq y_{(2)} \leq \cdots \leq y_{(n)}$ be the ordered y values in the sample, s_n. Let i_{1y}, i_{2y} and i_{3y} be the integers such that $y_{(i_{1y})} \leq Q_{1y} \leq y_{(i_{1y}+1)}$; $y_{(i_{2y})} \leq Q_{2y} \leq y_{(i_{2y}+1)}$ and $y_{(i_{3y})} \leq Q_{3y} \leq y_{(i_{3y}+1)}$. Let $p_{1y} = \frac{i_{1y}}{n}$, $p_{2y} = \frac{i_{2y}}{n}$; and $p_{3y} = \frac{i_{3y}}{n}$ be the proportion of the y values in a sample that are less than or equal to Q_{1y}, Q_{2y}, and Q_{3y}, respectively. Let i_{1x}, i_{2x} and i_{3x} be the integers such that $X_{(i_{1x})} \leq Q_{1x} \leq X_{(i_{1x}+1)}$, $X_{(i_{2x})} \leq Q_{2x} \leq X_{(i_{2x}+1)}$ and $X_{(i_{1x})} \leq Q_{1x} \leq X_{(i_{3x}+1)}$. Let $p_{1x} = \frac{i_{1x}}{n}$, $p_{2x} = \frac{i_{2x}}{n}$ and $p_{3x} = \frac{i_{3x}}{n}$ be the proportion of the x values in the sample that are less than or equal to Q_{1x}, Q_{2x} and Q_{3x}, respectively, so that Q_{1y}, Q_{2y} and Q_{3y} are approximately the sample p_{1y}, p_{2y} and $p_{3y}th$ quantiles $\hat{Q}_{1y}(p_{1y})$, $\hat{Q}_{2y}(p_{2y})$ and $\hat{Q}_{3y}(p_{3y})$. Unfortunately, the values of p_{1y}, p_{2y} and p_{3y} are unobservable because the values of Q_{1y}, Q_{2y} and Q_{3y} are unknown and thus need to be estimated from a given sample. Let $\hat{p}_{1y}$, $\hat{p}_{2y}$ and $\hat{p}_{3y}$ are the estimator of p_{1y}, p_{2y} and p_{3y}, respectively, then $\hat{Q}_{1y}(\hat{p}_{1y})$, $\hat{Q}_{2y}(\hat{p}_{2y})$ and $\hat{Q}_{3y}(\hat{p}_{3y})$ are the estimators of the parameters $Q_y(p_{1y})$, $Q_y(p_{2y})$ and $Q_y(p_{3y})$ respectively. The sample first quartile $\hat{Q}_{1y}$, the sample second quartile $\hat{Q}_{2y}$ and the sample third quartile $\hat{Q}_{3y}$ are the special cases of the estimators $\hat{Q}_y(\hat{p}_{1y})$, $\hat{Q}_y(\hat{p}_{2y})$ and $\hat{Q}_y(\hat{p}_{3y})$ with $\hat{p}_{1y} = \frac{1}{4}$; $\hat{p}_{2y} = \frac{1}{2}$; and $\hat{p}_{3y} = \frac{3}{4}$, respectively. One can say $\hat{Q}_{1y} = \hat{Q}_y\left(\frac{1}{4}\right)$; $\hat{Q}_{2y} = \hat{Q}_y\left(\frac{1}{2}\right)$ and $\hat{Q}_{3y} = \hat{Q}_y\left(\frac{3}{4}\right)$. Likewise let Q_{1x}, Q_{2x} and Q_{3x} are approximately the sample p_{1x}, p_{2x} and p_{3x} quartiles $\hat{Q}_x(p_{1x})$, $\hat{Q}_x(p_{2x})$ and $\hat{Q}_x(p_{3x})$, respectively. Let $\hat{p}_{1x}$, $\hat{p}_{2x}$ and $\hat{p}_{3x}$ be any estimator of p_{1x}, p_{2x} and p_{3x}, respectively, then $\hat{Q}_x(\hat{p}_{1x})$, $\hat{Q}_x(\hat{p}_{2x})$ and $\hat{Q}_x(\hat{p}_{3x})$ are natural estimators of the parameters $Q_x(p_{1x})$, $Q_x(p_{2x})$ and $Q_x(p_{3x})$ respectively. The sample quartiles $\hat{Q}_{1x}$, $\hat{Q}_{2x}$ and $\hat{Q}_{3x}$ are the special cases of the estimators $\hat{Q}_x(\hat{p}_{1x})$, $\hat{Q}_x(\hat{p}_{2x})$ and $\hat{Q}_x(\hat{p}_{3x})$ with $\hat{p}_{1x} = \frac{1}{4}$, $\hat{p}_{2x} = \frac{1}{2}$ and $\hat{p}_{3x} = \frac{3}{4}$ respectively. In other words, $\hat{Q}_{1x} = \hat{Q}_x\left(\frac{1}{4}\right)$, $\hat{Q}_{2x} = \hat{Q}_x\left(\frac{1}{2}\right)$ and $\hat{Q}_{3x} = \hat{Q}_x\left(\frac{3}{4}\right)$. Now consider the different proportions defined in Table A.1.

Table A.1 Population proportion of data values

	$X \leq Q_{1x}$	$X \leq Q_{2x}$	$X \leq Q_{3x}$	$Y \leq Q_{1y}$	$Y \leq Q_{2y}$	$Y \leq Q_{3y}$
$X \leq Q_{1x}$	0.25	0.25	0.25	$P_{Q_{1x}Q_{1y}}$	$P_{Q_{1x}Q_{2y}}$	$P_{Q_{1x}Q_{3y}}$
$X \leq Q_{2x}$	0.25	0.50	0.50	$P_{Q_{2x}Q_{1y}}$	$P_{Q_{2x}Q_{2y}}$	$P_{Q_{2x}Q_{3y}}$
$X \leq Q_{3x}$	0.25	0.50	0.75	$P_{Q_{3x}Q_{1y}}$	$P_{Q_{3x}Q_{2y}}$	$P_{Q_{3x}Q_{3y}}$
$Y \leq Q_{1y}$	$P_{Q_{1x}Q_{1y}}$	$P_{Q_{2x}Q_{1y}}$	$P_{Q_{3x}Q_{1y}}$	0.25	0.25	0.25
$Y \leq Q_{2y}$	$P_{Q_{1x}Q_{2y}}$	$P_{Q_{2x}Q_{2y}}$	$P_{Q_{3x}Q_{2y}}$	0.25	0.50	0.50
$Y \leq Q_{3y}$	$P_{Q_{1x}Q_{3y}}$	$P_{Q_{2x}Q_{3y}}$	$P_{Q_{3x}Q_{3y}}$	0.25	0.50	0.75

where,

$P_{Q_{1x}Q_{1y}} = P(X \leq Q_{1x} \cap Y \leq Q_{1y})$ be the population proportion of data values such that the value of the auxiliary variable X is less than or equal to Q_{1x} and the value of the study variable Y is less than or equal to Q_{1y}.

$P_{Q_{1x}Q_{2y}} = P(X \leq Q_{1x} \cap Y \leq Q_{2y})$ be the population proportion of data values such that the value of the auxiliary variable X is less than or equal to Q_{1x} and the value of the study variable Y is less than or equal to Q_{2y}.

$P_{Q_{1x}Q_{3y}} = P(X \leq Q_{1x} \cap Y \leq Q_{3y})$ be the population proportion of data values such that the value of the auxiliary variable X is less than or equal to Q_{1x} and the value of the study variable Y is less than or equal to Q_{3y}.

$P_{Q_{2x}Q_{1y}} = P(X \leq Q_{2x} \cap Y \leq Q_{1y})$ be the population proportion of data values such that the value of the auxiliary variable X is less than or equal to Q_{2x} and the value of the study variable Y is less than or equal to Q_{1y}.

$P_{Q_{2x}Q_{2y}} = P(X \leq Q_{2x} \cap Y \leq Q_{2y})$ be the population proportion of data values such that the value of the auxiliary variable X is less than or equal to Q_{2x} and the value of the study variable Y is less than or equal to Q_{2y}.

$P_{Q_{2x}Q_{3y}} = P(X \leq Q_{2x} \cap Y \leq Q_{3y})$ be the population proportion of data values such that the value of the auxiliary variable X is less than or equal to Q_{2x} and the value of the study variable Y is less than or equal to Q_{3y}.

$P_{Q_{3x}Q_{1y}} = P(X \leq Q_{3x} \cap Y \leq Q_{1y})$ be the population proportion of data values such that the value of the auxiliary variable X is less than or equal to Q_{3x} and the value of the study variable Y is less than or equal to Q_{1y}.

$P_{Q_{3x}Q_{2y}} = P(X \leq Q_{3x} \cap Y \leq Q_{2y})$ be the population proportion of data values such that the value of the auxiliary variable X is less than or equal to Q_{3x} and the value of the study variable Y is less than or equal to Q_{2y}.

$P_{Q_{3x}Q_{3y}} = P(X \leq Q_{3x} \cap Y \leq Q_{3y})$ be the population proportion of data values such that the value of the auxiliary variable X is less than or equal to Q_{3x} and the value of the study variable Y is less than or equal to Q_{3y}.

Let us define

$$\delta_{1x} = \frac{\hat{Q}_{1x}}{Q_{1x}} - 1, \quad \delta_{2x} = \frac{\hat{Q}_{2x}}{Q_{2x}} - 1, \quad \delta_{3x} = \frac{\hat{Q}_{3x}}{Q_{3x}} - 1$$

$$\delta_{1y} = \frac{\hat{Q}_{1y}}{Q_{1y}} - 1, \quad \delta_{2y} = \frac{\hat{Q}_{2y}}{Q_{2y}} - 1, \quad \delta_{3y} = \frac{\hat{Q}_{3y}}{Q_{3y}} - 1$$

such that

$$E\left(\delta_{1x}\right) \approx E\left(\delta_{2x}\right) \approx E\left(\delta_{3x}\right) \approx E\left(\delta_{1y}\right) \approx E\left(\delta_{2y}\right) \approx E\left(\delta_{3y}\right) \approx 0$$

$$E\left(\delta_{1x}^2\right) = \left(\frac{1-f}{n}\right)\left(\frac{3}{16}\right)\frac{1}{Q_{1x}^2 \{f_x(Q_{1x})\}^2},$$

$$E\left(\delta_{2x}^2\right) = \left(\frac{1-f}{n}\right)\left(\frac{1}{4}\right)\frac{1}{Q_{2x}^2 \{f_x(Q_{2x})\}^2}$$

$$E\left(\delta_{3x}^2\right) = \left(\frac{1-f}{n}\right)\left(\frac{3}{16}\right)\frac{1}{Q_{3x}^2 \{f_x(Q_{3x})\}^2},$$

$$E\left(\delta_{1y}^2\right) = \left(\frac{1-f}{n}\right)\left(\frac{3}{16}\right)\frac{1}{Q_{1y}^2 \{f_y(Q_{1y})\}^2}$$

$$E\left(\delta_{2y}^2\right) = \left(\frac{1-f}{n}\right)\left(\frac{1}{4}\right)\frac{1}{Q_{2y}^2 \{f_y(Q_{2y})\}^2},$$

$$E\left(\delta_{3y}^2\right) = \left(\frac{1-f}{n}\right)\left(\frac{3}{16}\right)\frac{1}{Q_{3y}^2 \{f_y(Q_{3y})\}^2}$$

$$E\left(\delta_{1x}\delta_{2x}\right) = \left(\frac{1-f}{8n}\right)\frac{1}{f_x(Q_{1x})f_x(Q_{2x})Q_{1x}Q_{2x}},$$

$$E\left(\delta_{1y}\delta_{2y}\right) = \left(\frac{1-f}{8n}\right)\frac{1}{f_y(Q_{1y})f_y(Q_{2y})Q_{1y}Q_{2y}}$$

$$E\left(\delta_{1x}\delta_{3x}\right) = \left(\frac{1-f}{16n}\right)\frac{1}{f_x(Q_{1x})f_x(Q_{3x})Q_{1x}Q_{3x}},$$

$$E\left(\delta_{1y}\delta_{3y}\right) = \left(\frac{1-f}{16n}\right)\frac{1}{f_y(Q_{1y})f_y(Q_{3y})Q_{1y}Q_{3y}}$$

$$E\left(\delta_{2x}\delta_{3x}\right) = \left(\frac{1-f}{8n}\right)\frac{1}{f_x(Q_{2x})f_x(Q_{3x})Q_{2x}Q_{3x}},$$

$$E\left(\delta_{2y}\delta_{3y}\right) = \left(\frac{1-f}{8n}\right)\frac{1}{f_y(Q_{2y})f_y(Q_{3y})Q_{2y}Q_{3y}}$$

$$E\left(\delta_{1x}\delta_{1y}\right) = \left(\frac{1-f}{n}\right)\frac{\{f_x(Q_{1x})\}^{-1}\{f_y(Q_{1y})\}^{-1}}{Q_{1x}Q_{1y}}\left(P_{Q_{1x}Q_{1y}} - \frac{1}{16}\right),$$

$$E\left(\delta_{2x}\delta_{1y}\right) = \left(\frac{1-f}{n}\right)\frac{\{f_x(Q_{2x})\}^{-1}\{f_y(Q_{1y})\}^{-1}}{Q_{2x}Q_{1y}}\left(P_{Q_{2x}Q_{1y}} - \frac{1}{8}\right),$$

$$E\left(\delta_{3x}\delta_{1y}\right) = \left(\frac{1-f}{n}\right)\frac{\{f_x(Q_{3x})\}^{-1}\{f_y(Q_{1y})\}^{-1}}{Q_{3x}Q_{1y}}\left(P_{Q_{3x}Q_{1y}} - \frac{3}{16}\right),$$

$$E\left(\delta_{1x}\delta_{2y}\right) = \left(\frac{1-f}{n}\right)\frac{\{f_x(Q_{1x})\}^{-1}\{f_y(Q_{2y})\}^{-1}}{Q_{1x}Q_{2y}}\left(P_{Q_{1x}Q_{2y}} - \frac{1}{8}\right),$$

$$E\left(\delta_{2x}\delta_{2y}\right) = \left(\frac{1-f}{n}\right)\frac{\{f_x(Q_{2x})\}^{-1}\{f_y(Q_{2y})\}^{-1}}{Q_{2x}Q_{2y}}\left(P_{Q_{2x}Q_{2y}} - \frac{1}{4}\right),$$

$$E\left(\delta_{3x}\delta_{2y}\right) = \left(\frac{1-f}{n}\right)\frac{\{f_x(Q_{3x})\}^{-1}\{f_y(Q_{2y})\}^{-1}}{Q_{3x}Q_{2y}}\left(P_{Q_{3x}Q_{2y}} - \frac{3}{8}\right),$$

$$E\left(\delta_{1x}\delta_{3y}\right) = \left(\frac{1-f}{n}\right)\frac{\{f_x(Q_{1x})\}^{-1}\{f_y(Q_{3y})\}^{-1}}{Q_{1x}Q_{3y}}\left(P_{Q_{1x}Q_{3y}} - \frac{3}{16}\right),$$

$$E\left(\delta_{2x}\delta_{3y}\right) = \left(\frac{1-f}{n}\right)\frac{\{f_x(Q_{2x})\}^{-1}\{f_y(Q_{3y})\}^{-1}}{Q_{2x}Q_{3y}}\left(P_{Q_{2x}Q_{3y}} - \frac{3}{8}\right),$$

and

$$E\left(\delta_{3x}\delta_{3y}\right) = \left(\frac{1-f}{n}\right)\frac{\{f_x(Q_{3x})\}^{-1}\{f_y(Q_{3y})\}^{-1}}{Q_{3x}Q_{3y}}\left(P_{Q_{3x}Q_{3y}} - \frac{9}{16}\right),$$

These approximate expected values are obtained by following Kuk and Mak [9] or can be had from Singh, Solanki, and Singh [10], Singh, Puertas and Singh [11], and Eappen, Sedory and Singh [5,6].

A.2 SAS CODES USED IN THE SIMULATION STUDY

```
*SAS codes used for the simulation;
```

```
%letns_in=2500;
data data1;
np=10000;
ay=1.5;
by=2.0;
ax=1.6;
bx=1.8;
callstreaminit(12345);
do I=1to np by1;
ystar= rand('gamma',ay,by);
xstar= rand('gamma',ax,bx);
sigmays= sqrt(ay*by**2);
meanys= ay*by;
sigmaxs= sqrt(ax*bx**2);
meanxs= ax*bx;
output;
end;
data data2;
set data1;
do rho= 0.8to0.98by0.02;
meany= 11;
meanx= 15;
yp= meany + (ystar-meanys)*sqrt(1-rho**2) +
sigmays*rho*(xstar-meanxs)/sigmaxs;
xp= meanx+(xstar-meanxs);
ypxp = yp*xp;
yp2 = yp*yp;
xp2 = xp*xp;
output;
end;
keep rho ypxpypypxp yp2 xp2;
procsortdata=data2;
by rho;
PROCMEANSDATA=DATA2 NOPRINT;
BY rho;
VARypxpypypxp yp2 xp2;
OUTPUTOUT = DATA201 SUM= SYP SXP SYPXP SYP2 SXP2 N=NP;
```

```
DATA DATA202;
SET DATA201;
RHOXY = (NP*SYPXP-SYP*SXP)/(SQRT(NP*SYP2-
SYP**2)*SQRT(NP*SXP2-SXP**2));
KEEP RHO RHOXY;
PROCPRINTDATA=DATA202;
run;
procprintdata=data2;
run;
procunivariatedata=data2 noprint;
by rho;
varyp;
histogramyp/gamma;
odsoutputparameterestimates = data3y;
run;
procprintdata = data3y;
run;
data data4y (keep = by_e rho) data5y (keep=ay_e rho);
set data3y;
if symbol = 'Sigma'thendo; by_e = estimate;output
data4y;end;
if symbol = 'Alpha'thendo; ay_e = estimate;output
data5y;end;
procprintdata=data4y;
procprintdata=data5y;
run;
odstraceon;
odsoutputparameterestimates = data3x;
procunivariatedata=data2;
by rho;
varxp;
histogramxp/gamma;
run;
odstraceoff;
*proc print data = data3x;
run;
data data4x (keep = bx_e rho) data5x (keep=ax_e rho);
set data3x;
if symbol = 'Sigma'thendo; bx_e = estimate;output
data4x;end;
if symbol = 'Alpha'thendo; ax_e = estimate;output
data5x;end;
procprintdata=data4x;
procprintdata=data5x;
run;
data data6;
merge data4y data5y data4x data5x;
procprintdata=data6;
run;
procmeansdata=data2 noprint;
```

```
varypxp;
by rho;
outputout=data7 q1=q1yp q1xp median= q2yp q2xp q3=q3yp
q3xp;
run;
data data8;
merge data6 data7;
by rho;
drop _type_;
procprintdata=data7;
run;
*data8 is quartiles;
procprintdata=data8;
run;
data data9;
set data8;
fxq1x= (q1xp**(ax_e-1)*exp(-q1xp/bx_e))/(gamma
(ax_e)*bx_e**ax_e);
fyq1y= (q1yp**(ay_e-1)*exp(-q1yp/by_e))/(gamma
(ay_e)*by_e**ay_e);
fxq2x= (q2xp**(ax_e-1)*exp(-q2xp/bx_e))/(gamma
(ax_e)*bx_e**ax_e);
fyq2y= (q2yp**(ay_e-1)*exp(-q2yp/by_e))/(gamma
(ay_e)*by_e**ay_e);
fxq3x= (q3xp**(ax_e-1)*exp(-q3xp/bx_e))/(gamma
(ax_e)*bx_e**ax_e);
fyq3y= (q3yp**(ay_e-1)*exp(-q3yp/by_e))/(gamma
(ay_e)*by_e**ay_e);
procprintdata=data9;
run;
procsortdata=data2;
by rho;
procsortdata=data7;
by rho;
data data10;
merge data2(in=abc) data7 (in=cde);
by rho;
ifabc and cde;
*proc print data=data10;
run;
data data11;
set data10;
ifyp le q1yp then iyq1=1; else iyq1=0;
ifxp le q1xp then ixq1=1; else ixq1=0;
ifyp le q2yp then iyq2=1; else iyq2=0;
ifxp le q2xp then ixq2=1; else ixq2=0;
ifyp le q3yp then iyq3=1; else iyq3=0;
ifxp le q3xp then ixq3=1; else ixq3=0;
pq1xq1y = ixq1*iyq1;
pq1xq2y = ixq1*iyq2;
```

```
pq1xq3y = ixq1*iyq3;
pq2xq1y = ixq2*iyq1;
pq2xq2y = ixq2*iyq2;
pq2xq3y = ixq2*iyq3;
pq3xq1y = ixq3*iyq1;
pq3xq2y = ixq3*iyq2;
pq3xq3y = ixq3*iyq3;
keep rho ypxp iyq1 ixq1 iyq2 ixq2 iyq3 ixq3 pq1xq1y pq1xq2y
pq1xq3y pq2xq1y pq2xq2y pq2xq3y pq3xq1y pq3xq2y pq3xq3y;
*proc print data=data11;
run;
procmeansdata=data11 noprint;
var pq1xq1y pq1xq2y pq1xq3y pq2xq1y pq2xq2y pq2xq3y pq3xq1y
pq3xq2y pq3xq3y;
by rho;
outputout=data12 mean = pq1xq1y pq1xq2y pq1xq3y pq2xq1y
pq2xq2y pq2xq3y pq3xq1y pq3xq2y pq3xq3y;
procprintdata=data12;
var rho pq1xq1y pq1xq2y pq1xq3y pq2xq1y pq2xq2y pq2xq3y
pq3xq1y pq3xq2y pq3xq3y;
run;
data data13;
merge data9 data12 data202;
by rho;
drop _type_ ypxp;
rename _freq_ = np;
procprintdata =data13;
run;
data data14;
set data13;
ns=&ns_in;
f=ns/np;
fact=(1-f)/ns;
*empirical mode****;
moy = 3*q2yp-q1yp-q3yp;
mox = 3*q2xp-q1xp-q3xp;
v1 = 9/(4*fyq2y**2);
v2 = 3/(16*fyq1y**2);
v3 = 3/(16*fyq3y**2);
v4 = -3/(4*fyq1y*fyq2y);
v5 = -3/(4*fyq2y*fyq3y);
v6 = 1/(8*fyq1y*fyq3y);
vmoy = fact*(v1+v2+v3+v4+v5+v6);
********bias of the ratio of moy*****;
b1 = 9/(4*fxq2x**2);
b2 = 3/(16*fxq1x**2);
b3 = 3/(16*fxq3x**2);
b4 = -3/(4*fxq1x*fxq2x);
b5 = -3/(4*fxq2x*fxq3x);
b6 = 1/(8*fxq1x*fxq3x);
```

```
b7 = 9*(pq2xq2y-1/4)/(fxq2x*fyq2y);
b8 = -3*(pq2xq1y-1/8)/(fxq2x*fyq1y);
b9 = -3*(pq2xq3y-3/8)/(fxq2x*fyq3y);
b10 = -3*(pq1xq2y-1/8)/(fxq1x*fyq2y);
b11 = (pq1xq1y-1/16)/(fxq1x*fyq1y);
b12 = (pq1xq3y-3/16)/(fxq1x*fyq3y);
b13 = -3*(pq3xq2y-3/8)/(fxq3x*fyq2y);
b14 = (pq3xq1y-3/16)/(fxq3x*fyq1y);
b15 = (pq3xq3y-9/16)/(fxq3x*fyq3y);
brmoy= (moy/mox**2)*fact*(b1+b2+b3+b4+b5+b6)-(1/
mox)*fact*(b7+b8+b9+b10+b11+b12+b13+b14+b15);
rb_br= brmoy*100/ moy;
**************mse of ratio type estimator******;
mse_ratio=fact*(v1+v2+v3+v4+v5+v6)+fact*(moy/
mox)**2*(b1+b2+b3+b4+b5+b6)-2*fact*(moy/
mox)*(b7+b8+b9+b10+b11+b12+b13+b14+b15);
re_ratio= vmoy*100/mse_ratio;
********regression type estimator******;
b = (b7+b8+b9+b10+b11+b12+b13+b14+b15)/
(b1+b2+b3+b4+b5+b6);
vreg=fact*(v1+v2+v3+v4+v5+v6)
+fact*b**2*(b1+b2+b3+b4+b5+b6)-
2*b*fact*(b7+b8+b9+b10+b11+b12+b13+b14+b15);
A=v1+v2+v3+v4+v5+v6;
a2= b7+b8+b9+b10+b11+b12+b13+b14+b15;
a3=b1+b2+b3+b4+b5+b6;
min_mse=fact*A-fact*(a2**2/a3);
re_reg= vmoy*100/vreg;
keep ns rho rhoxyrb_brre_ratiore_reg;
procprintdata=data14;
run;
```

Chapter 5

On the distribution of a busy period for the single server queue with balking, catastrophes and repairs

Sherif I. Ammar

5.1 INTRODUCTION

In recent years there has been renewed interest in discussing the behavior of queueing models with catastrophes [1]. The notion of catastrophes occurring at random, leading to annihilation of all customers there and the momentary inactivation of the service facilities until a new customer arrives is not uncommon in many practical situations. This can be seen as negative customers to the system and their characteristic is to remove some or all the regular customers in the system. The catastrophes may come either from outside the system or from another service station. For example, in computer networks, if a job infected with a virus arrives, it transmits the virus to other processors and inactivates them [2]. Hence, computer networks with a virus infection may be modeled by queueing networkers with catastrophes. Other interesting articles in the area include [3–5].

Queueing systems with a repairable server often arise in [6–8]. Such repairable server queueing models are interesting, either from the point of view of queueing theory or of reliability. These phenomena occur in the area of computer and communication systems where failure and repair of processers have a major impact on the flow of jobs that have to be handled by the processors [9,10].

Another salient feature, which has been widely studied in the literature, is queueing systems subject to balking. Balking means that arrival customers might decide to balk (not to join the system) when they find too many customers lined up in the system. Balking is not only a common phenomenon in queues arising in daily activities but also finds application in various machine interference/models e.g., see [11–13]. Examples of balking can be found in many situations involving critical patients, communication systems and service systems such as telephone switchboard system and perishable goods storage inventory system.

On the other hand, the busy period analysis of queueing systems is an integral part of a queueing system as the distribution of its duration is important from the server's point of view and is also helpful in the efficient

DOI: 10.1201/9781003356653-5

planning of the system and resources. In particular, the busy period plays an important role in understanding various operations taking place in any queueing system and helps to improve the management of systems.

The busy period for the Poisson queue has been studied by Plam [14,15] and later followed among others by Takcacs [16], Neuts [17], Erlander [18], Karlin and McGregor [19], Rice [20], Nance et al. [21], Shanbhag [22], Conolly [23,24], Bunday and El-Badri [25], Parthasarthy and Sharafali [26], Kumar [27], Sharma [28], Tarabia [29] and Stadje [30].

In computer systems, if there are many customers (data) lined up in the system, a new arrival may decide not to join the system. In addition, if data is infected with virus, it may annihilate or transmit to other processors. These systems can be represented as queueing models with balking, catastrophes and repairs. These queueing models have wide applications in telecommunication systems and manufacturing systems, etc. Consequently, many researchers have studied queueing systems with catastrophes and balking (see [31–36]). Also, Ammar [37] has discussed the transient behavior of a two-processor heterogeneous system with catastrophes, server failures and repairs, and Tarabia [38] has investigated the combined effects of catastrophes and balking.

As we mentioned earlier, there are several papers dealing with both phenomena of catastrophes and balking. But no work is found in the literature which studied the distribution of the busy period taking the above-mentioned features together. Based on the literature review, we have adopted the procedures of Ammar [37], for the transient state of Markovian queue to obtain the transient behavior of the distribution of busy period for the Markovian single server queue with balking, catastrophes and repairs where balking occurs if and only if the system size equals or exceeds a threshold value k.

This paper is organized as follows: Section 5.2 gives a description of busy period with balking and reneging and a simple equation is presented to obtain the density function of the length of a busy period for an M/M/1 queue with balking and catastrophes and repairs. In Section 5.3, we perform sensitivity analysis through numerical experiments. Finally, the main conclusion of the paper is summarized in section 5.4.

5.2 MODEL DESCRIPTION AND BUSY PERIOD ANALYSIS

The arrival process of a customer is a Poisson process with a mean arrival rate λ during times that the server is working. Assume that the service discipline is FCFS with the service time following an exponential distribution with mean $1/\mu$. An arriving customer will decide to join the queue with probability one if the number of customers in the system is less than a threshold value k. If there are k customers or more ahead of him, then he joins the queue with probability p and may balk with probability $1 - p$. Apart from arrival and service processes, when the system is not empty, catastrophes also occur at the service facility according to a Poisson process with rate γ. Whenever a catastrophe occurs at the operational server, all customers in the system are flushed out immediately and the server gets inactivated i.e., the server is subject to failure.

The repair times of a failed server are i.i.d, according to an exponential distribution with mean $1/\eta$. After repairing the server becomes ready to serve new customers. Let $Q(t)$ be the probability that the server is under repair at time t, and $P_n(t) = P\{n$ customers in the system at $t\}$. We assume that there is one customer in the system at $t = 0$. Therefore, we postulate that $P_n(0) = \delta_{1n}$ where δ_{1n} is the Kronecker delta and the state $n = 0$ is absorbing.

In view of these assumptions and by the forward Kolmogorov equations, the state probabilities and the failure probability can be described by the differential-difference equations as follows:

$$\frac{dQ(t)}{dt} = -\eta Q(t) + \gamma\,[1 - Q(t)] \tag{5.1}$$

$$\frac{dP_0(t)}{dt} = \mu P_1(t) + \eta Q(t), \tag{5.2}$$

$$\frac{dP_1(t)}{dt} = -(\lambda + \mu + \gamma)P_1(t) + \mu P_2(t), \tag{5.3}$$

$$\frac{dP_n(t)}{dt} = -(\lambda + \mu + \gamma)P_n(t) + \lambda P_{n-1}(t) + \mu P_{n+1}(t), \quad 2 \leq n \leq k - 1 \tag{5.4}$$

$$\frac{dP_k(t)}{dt} = -(\lambda_p + \mu + \gamma)P_k(t) + \lambda P_{k-1}(t) + \mu P_{k+1}(t), \tag{5.5}$$

$$\frac{dP_n(t)}{dt} = -(\lambda_p + \gamma + \mu)P_n(t) + \lambda_p P_{n-1}(t) + \mu P_{n+1}(t) \quad n \geq k + 1 \tag{5.6}$$

If $f(t)$ is the probability density function of the length of the busy period, then evidently $f(t) = \frac{dP_0(t)}{dt}$.

5.2.1 Evaluation for $P_k(t)$

In the analysis below, we will express $P_k(t)$, $\quad k \geq 1$, in explicit form in terms of the modified Bessel functions.

We define generating function $P_k(z, t)$, or simply $P(z, t)$ for the number of customers awaiting commencement of service as

$$P(s, t) = Q(t) + q_k(t) + + \sum_{n=k+1}^{\infty} P_n(t) z^{n-k} \quad P(z, 0) = 1 \text{ and } Q(0) = 0$$

(5.7)

with

$$q_k(t) = \sum_{n=0}^{k} P_n(t)$$

Adding equations (5.1)–(5.5) from the above system, we have

$$Q'(t) + R'_k(t) = \gamma [1 - Q(t)] - \gamma q_k(t) - \lambda_p P_k(t) + \mu P_{k+1}(t)$$

(5.8)

Using the previous equation (5.8) and the definition of $P(s, t)$, we get

$$\frac{\partial P(z, t)}{\partial t} - \left[\lambda_p z - (\lambda_p + \gamma + \mu) + \frac{\mu}{z} \right] P(z, t)$$

$$= \left[\lambda_p s - (\lambda_p + \gamma + \mu) + \frac{\mu}{z} \right] \left[Q(t) + q_k(t) \right] + \lambda_p (z - 1) P_k(t) + \gamma \quad (5.9)$$

Solving the first order differential equation (5.9) using the integrating factor $\exp \left\{ -\left[\lambda_p z - (\lambda_p + \gamma + \mu) + \frac{\mu}{z} \right] t \right\}$, we have

$$P(z, t) = \exp \left\{ -\left[\lambda_p z - (\lambda_p + \gamma + \mu) + \frac{\mu}{z} \right] t \right\}$$

$$+ \int_0^t \left[\lambda_p (z - 1) P_k(u) + \left\{ \lambda_p z - (\lambda_p + \gamma + \mu) + \frac{\mu}{z} \right\} (Q(u) + q_k(u)) + \gamma \right]$$

$$\cdot \exp \left\{ -\left[\lambda_p z - (\lambda_p + \gamma + \mu) + \frac{\mu}{z} \right] (t - u) \right\} du.$$

(5.10)

Rewriting to

$$\lambda_p z - (\lambda_p + \gamma + \mu) + \frac{\mu}{z} = -a + bs + \frac{c}{z}$$

where $a = (\lambda_p + \gamma + \mu)$, $b = \lambda_p$ and $c = \mu$.

On account of $\alpha = 2\sqrt{bc}$, $\beta = \sqrt{\frac{b}{c}}$ and using Bessel function properties, we get

$$\exp\left\{\left[-a + bz + \frac{c}{z}\right]t\right\} = \exp\{-at\}\exp\left\{\left[bs + \frac{c}{z}\right]t\right\}$$

$$= \exp\{-at\} \sum_{n=-\infty}^{\infty} (\beta z)^n I_n(\alpha t)$$

and

$$P(z, t) = \exp\{-at\} \sum_{n=-\infty}^{\infty} (\beta z)^n I_n(\alpha t)$$

$$+ \int_0^t \exp\{-a(t - u)\}\left[\lambda_p(z - 1)P_k(u) + \left\{bz - a + \frac{c}{z}\right\}(Q(u)\right.$$

$$\left. + q_k(u)) + \gamma\right] \cdot \sum_{n=-\infty}^{\infty} (\beta z)^n I_n(\alpha(t - u))\,du.$$

$$(5.11)$$

Comparing the coefficient of s^n on either side and solving for $Q(t) + q_k(t)$, we get

$$Q(u) + q_k(u) = \exp(-at)\, I_0(\alpha t)$$

$$+ \lambda_p \int_0^t \exp(-a(t - u))[\{\beta^{-1}I_1(\alpha(t - u))$$

$$- I_0(\alpha(t - u))\}P_k(u) + \{(\lambda_p + \mu)I_0(\alpha(t - u)) \quad (5.12)$$

$$- 2\lambda_p(I_1(\alpha(t - u)))\}(Q(u) + q_k(u))$$

$$- \gamma I_0(\alpha(t - u)]\,du.$$

where $Q(t)$ is obtained from (5.1) by

$$Q(t) = \frac{\gamma}{\gamma + \eta}[1 - e^{-(\gamma+\eta)t}] \qquad (5.13)$$

In the sequel, for any function $g(t)$, $g_k^*(s)$ is its Laplace transform. Now taking the Laplace transform and solving the above equation for $Q^*(s) + q_k^*(s)$, we have

$$(\gamma + s)(Q^*(s) + q_k^*(s)) = 1 + \frac{1}{2}\left[(s + a) - \sqrt{(s + a)^2 - \alpha^2} - \alpha\beta\right]P_k^*(s) + \frac{\gamma}{s} \quad (5.14)$$

Now, in order to solve for $P_1(t)$, we consider the system of equations (5.2)–(5.4) subject to the condition (5.14). The system (5.4) together with (5.2) and (5.3) can be expressed in the form

$$\frac{d\mathbf{P}(t)}{dt} = \mathbf{AP(t)} + \eta e_1 Q(t) + \mu P_k(t)e_k \tag{5.15}$$

where $\mathbf{A} = (a_{ij})_{k \times k}$ is given as:

$$a_{ij} = \begin{cases} \lambda, & j = i - 1, \quad i = 2, 3, \dots, k - 1, \\ -(\lambda + \mu + \gamma), & j = i, \qquad i = 1, 2, \dots, k - 1, \\ \mu, & j = i + 1, \quad i = 0, 1, 2, \dots, k - 2, \end{cases}$$

$P(t) = (P_0(t), P_1(t), \dots\dots\dots, P_{k-1}(t))^T$ and e_k is a column vector of order k with 1 in the k-th place and the remaining elements are zero.

Taking Laplace transform the solution of (5.12) is

$$\hat{\mathbf{P}}(s) = [sI - A]^{-1}[\eta e_1 Q(s) + \mu \hat{P}_k(s)e_k + \mathbf{P}(0)] \tag{5.16}$$

with

$$\mathbf{P}(0) = e_2 = (0, 1, \dots, 0)^T_{k \times 1}.$$

Now $[sI - A]$ can be partitioned as

$$[sI - A] = \begin{bmatrix} s & -\mu E_1^T \\ 0 & B_{k-1}(s) \end{bmatrix}$$

where $\mathbf{E}_i$ is a column vector of order $k - 1$ with one in the i-th place and zero in the rest of the places and $B_{k-1}(s)$ is of order $(k - 1) \times (k - 1)$

Hence

$$[sI - A]^{-1} = \begin{bmatrix} \frac{1}{s} & -\frac{\mu}{s}E_1^T B_{k-1}^{-1}(s) \\ 0 & B_{k-1}^{-1}(s) \end{bmatrix} \tag{5.17}$$

Let

$$\mathbf{B}_{k-1}^{-1}(s) = (b^*_{i+1, j+1}(s))_{(k-1) \times (k-1)}.$$

Consider $\mathbf{B}_{k-1}(s)$ as an almost lower triangular matrix and we adopt the procedure for finding the inverse of such a matrix by following the technique used in [39], (see Appendix I).

Substituting in (5.14) and using (5.16), we obtain after simplification,

$$P_k^*(s) = \cfrac{\begin{aligned} &1 - (s + \gamma)(s + \mu)b_{1,1}^*(s) - (s + \gamma)\sum_{i=2}^{k-1} s b_{i,1}(s) \\ &- [(s + \gamma) + \eta(1 + b_{1,1}^*(s))]Q^*(s) + \tfrac{\gamma}{s} \end{aligned}}{\begin{aligned} &s + \lambda_p + \gamma - \tfrac{1}{2}\left((s + a) - \sqrt{(s + a)^2 - \alpha^2}\right) \\ &+ \mu(s + \gamma)\left[(s + \mu)b_{1,1}^*(z) - \sum_{i=2}^{k-1} s b_{i,1}(s)\right] \end{aligned}} \tag{5.18}$$

It is clear that $b_{i,j}^*(s)$ are all rational algebraic functions in s. In particular, the cofactor of the (i, j)th element of $\mathbf{B}_{k-1}^{-1}(s)$ is a polynomial of degree $k - 1 - |i - j|$. When $i = j$, the cofactors are polynomials of degree $k - 1$ with the leading coefficient equal to 1. Also, the determiner of $\mathbf{B}_{k-1}(s)$ is a polynomial of degree k with leading coefficient equal to 1. In fact, $u_{k,1}(s) = 0$ is the characteristic equation of $\mathbf{B}_{k-1}(s)$. Since the element $b_{11}^*(s)$ of $\mathbf{B}_{k-1}(s)$ is non zero, it is also known that the characteristic roots of $\mathbf{B}_k(z)$ are all distinct and negative [40]. Hence, the inverse transform $b_{ij}(t)$ of $b_{ij}^*(s)$ can be obtained by partial fraction decomposition.

Let $z_i, i = 1, 2, \ldots, k - 1$ be the characteristic roots of $\mathbf{B}_{k-1}(s)$.

Then, after partial fraction decomposition, and simplification, (5.18) becomes

$$P_k^*(s) = \cfrac{R_{k-1}^*(s) - [(s + \gamma) + \eta(1 + b_{1,1}^*(s))]Q^*(s) + \tfrac{\gamma}{s}}{\tfrac{1}{2}\left((s + a) + \sqrt{(s + a)^2 - \alpha^2}\right)\left\{1 - \dfrac{2\mu}{(s + a) + \sqrt{(s + a)^2 - \alpha^2}}(1 - M_{k-1}^*(s))\right\}} \tag{5.19}$$

where

$$R_{k-1}^*(s) = \sum_{i=1}^{k-1} \frac{N_i}{s - s_i} \tag{5.20}$$

and

$$M_{k-1}^*(s) = \sum_{i=1}^{k-1} \frac{D_i}{s - s_i} \tag{5.21}$$

With the constants N_i and D_i given by

$$N_i = \lim_{s \to s_i}(s - s_i)\left[1 - (s + \mu)(s + \gamma)b_{1,1}^*(s) - (s + \gamma)\sum_{i=2}^{k-1} s b_{i,1}(s)\right] \tag{5.22}$$

and

$$D_i = \lim_{s \to s_i} (s - s_i) \left[(s + \mu)(s + \gamma) b_{i,1}^*(s) - (s + \gamma) \sum_{i=1}^{k-1} s b_{i,1}(s) \right] \qquad (5.23)$$

We note that

$$\left| \frac{2\mu}{(s + a) + \sqrt{(s + a)^2 - \alpha^2}} \{ 1 - M_{k-1}^*(s) \} \right| < 1, \text{ (see Appendix II)}.$$

Hence (5.19) simplifies to

$$P_k^*(s) = \left(\frac{2}{\alpha} \right) \left\{ R_{k-1}^*(s) - [(s + \gamma) + \eta(1 + b_{1,1}^*(s))] Q^*(s) + \frac{\gamma}{s} \right\}$$

$$\times \sum_{r=0}^{\infty} \sum_{m=0}^{r} (-1)^m \binom{r}{m} \left(\sqrt{\frac{\mu}{\lambda_p}} \right)^r \left[\frac{(s + a) + \sqrt{(s + a)^2 - \alpha^2}}{\alpha} \right]^{r+1} (M_{k-1}^*(s))^m$$

$$(5.24)$$

Which on inversion yields

$$P_k(t) = \sum_{r=0}^{\infty} \sum_{m=0}^{r} (-1)^m \binom{r}{m} \left(\sqrt{\frac{\mu}{\lambda_p}} \right)^r \left[\int_0^t R_{k-1}(t - u) \int_0^u M_{k-1}^{(c)(m)}(u - v) [I_r(rv) \right.$$

$$- I_{r+2}(rv)] dv du - \frac{r+1}{\mu \beta^{r+1}} \int_0^t M_{k-1}^{(c)(m)}(t - u) e^{-(\lambda_p + \gamma + \mu)u} Q(t)$$

$$\times (u + \gamma + \eta) \frac{I_{r+1}(\alpha u)}{u} du - \frac{r+1}{\mu \beta^{r+1}} \int_0^t \left[b_{11}(t - u) Q(t) \right.$$

$$\left. \times \int_0^u M_{k-1}^{(c)(m)}(u - v) e^{-(\lambda_p + \gamma + \mu)u} \frac{I_{r+1}(\alpha v)}{v} dv \right] du \right]$$

$$(5.25)$$

where $M^{(c)(m)}(t)$ is the m-fold convolution of $M(t)$ with itself with $M^{(c)(0)}(t) = \delta(t)$.

5.2.2 Evaluation for the busy density function

Now, since $P_1^*(s) = e_2^T P(s)$, we obtain from (5.16) as

$$\begin{aligned} P_1^*(s) &= b_{11}^*(s) + \eta b_{11}(s) Q(s) + \mu b_{1k-1}^*(s) P_k^*(s) \\ P_1^*(z) &= b_{11}^*(z) + \mu b_{1k-1}^*(z) P_k^*(z) \end{aligned} \qquad (5.26)$$

Which on inversion yields, for $k = 2, 3, \dots$,

$$P_1(t) = b_{11}(t) + \eta \int_0^t b_{11}(t-u)Q(u) + \mu \int_0^t b_{1k-1}(t-u)P_k(u)\,du \quad (5.27)$$

where $P_k(t)$ is given by (5.23).

Thus, the probability density function $f(t)$ of busy period is

$$
\begin{aligned}
f(t) &= \mu P_1(t) + \eta Q(t) \\
&= \frac{\gamma\eta}{\gamma+\eta}[1 - e^{-(\gamma+\eta)t}] + \mu \Big\{ b_{11}(t) + \eta \int_0^t b_{11}(t-u)Q(u) \\
&\quad + \mu \int_0^t b_{1k-1}(t-u)P_k(u)\,du \Big\}
\end{aligned}
$$

5.3 NUMERICAL ILLUSTRATIONS

In the previous section, we obtained an explicit expression for the busy period density function. In this section, we present some numerical illustrations to highlight the effect of the catastrophe rate γ, the repair rate η, the arrival rate λ, and the balking rate p on the behavior of the busy period distribution of the system. In all the examples we have taken a threshold value, $k = 1$ and the service rate, $\mu = 1$.

The graphs displayed in Figure 5.1 show the evolution of the busy period distribution function against time t with a variation of the catastrophe rate, γ. So, the numerical calculation, corresponding to three different values of γ are archived for the sake of comparison.

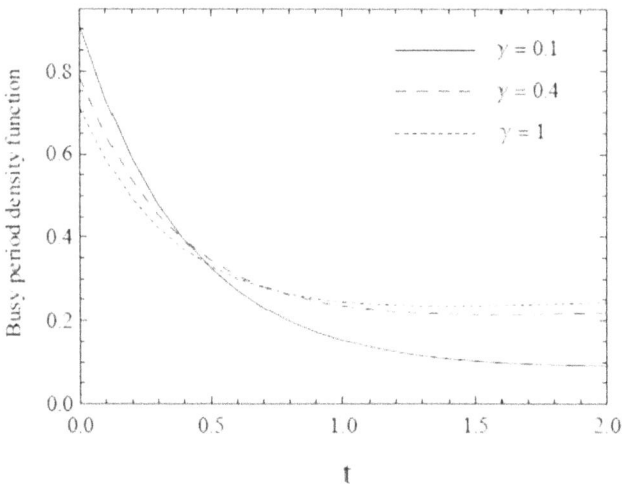

Figure 5.1 $\lambda = 0.3$, $p = 0.2$ and $\eta = 0.5$.

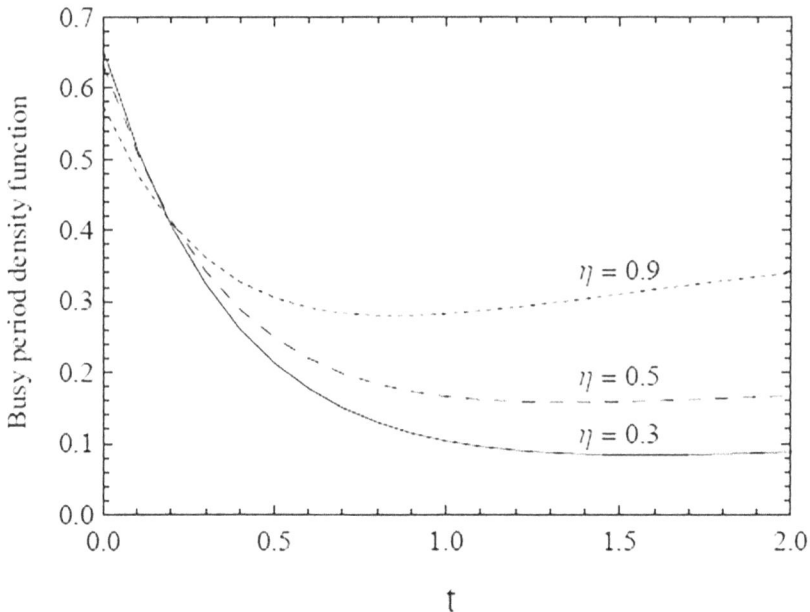

Figure 5.2 $\lambda = 0.7$, $p = 0.4$ and $\gamma = 0.5$.

It is clear at small values t, $(t < 0.5)$ that the increase in the values of γ results in a decrease in the values of $f(t)$. In contrast, at the relatively large values of t, $(t > 0.5)$ the values of $f(t)$. continuously increase with the increase of γ. This clearly indicates that the effect of the parameter γ on the behavior of the busy period distribution function will strongly depend on time t.

Figure 5.2 clarifies that the busy period function with changes of η behaves in a similar manner to that observed due to the changes of γ.

Figure 5.3 illustrates the variation of $f(t)$ as a function in time, t for different values of balking rate, p. It is observed that the busy period distribution function continuously decreases when the time goes on. In contrast, the values of $f(t)$ grow with the monotonic increase in the values of the parameter p.

In Figure 5.4, we aim to investigate the effect of the arrival rate, λ on the busy period behavior. It is shown that the variation of λ has a slight effect on $f(t)$, especially at large values of t.

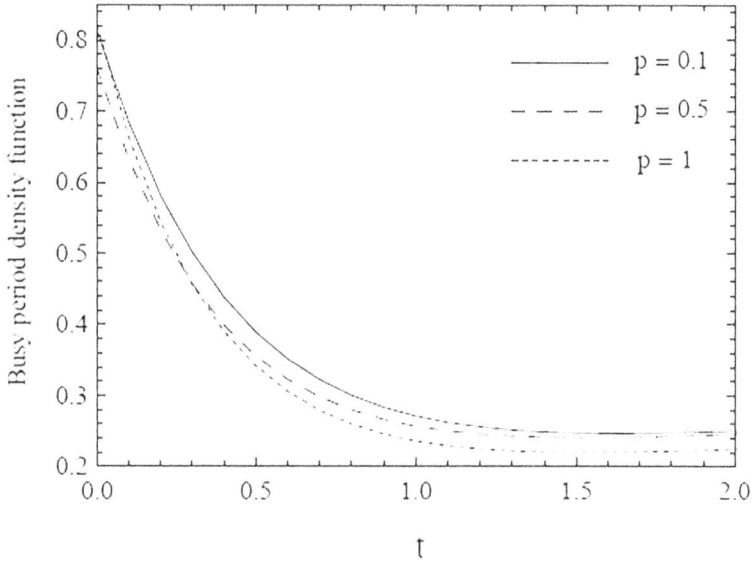

Figure 5.3 $\lambda = 0.4$, $\gamma = 0.3$ and $\eta = 0.5$.

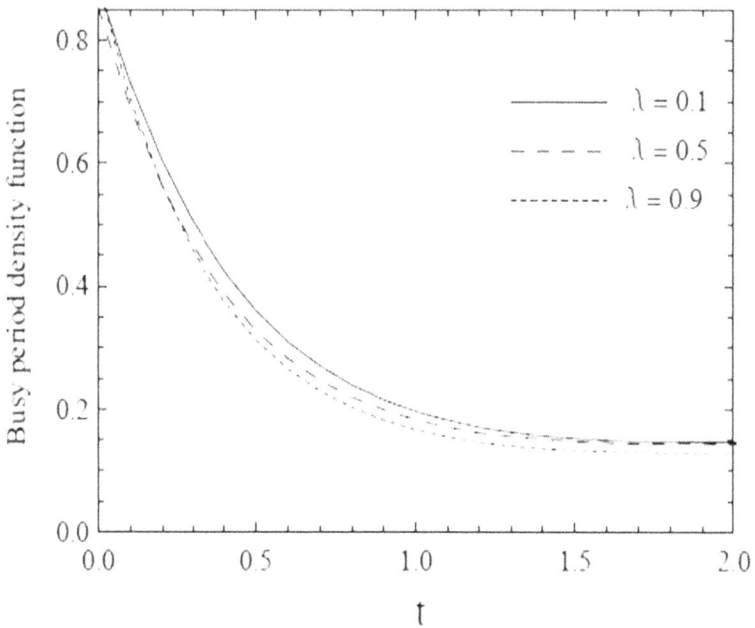

Figure 5.4 $p = 0.5$, $\gamma = 0.3$ and $\eta = 0.7$.

5.4 CONCLUSION

In this paper, we have investigated the transient behavior of the distribution of a busy period for an M/M/1 queue with balking, catastrophes and a repairable server. We have obtained an explicit expression of the busy period distribution function in terms of modified Bessel functions. Some numerical examples are presented to clarify the behavior of the busy period distribution function. Having obtained the results, some performance measures such as reliability, availability and the mean time to system failure of the system can be obtained.

REFERENCES

1. Gelenbe E. and Punjolle G., 1998. Introduction to Queueing Networks, Wiley Chichester.
2. Chao X., Miyazawa M., and Pinedo M., 1993. Queueing Networks – Customers, Signals, and Product form Solutions, Wiley Chichester.
3. Harrison P. G. and Patel N. M., 1993. Performance Modelling of Communication Networks, Wiley Chichester.
4. Henderson W., 1993. Queueing networks with negative customers and negative queue lengths, J. Appl. Probab. 30, 931–942.
5. Jain G. and Sigman K., 1996. A Pollaczek-Khinchine formula for M/G/1 queues with disasters, J. Appl. Probab. 33, 1191–1200.
6. Avi-Itzhak, B. and Naor, P., 1963. Some queueing problems with the service station subject to breakdowns, Oper. Res. 11, 303–320.
7. Neuts M.F. and Lucantoni D.M., 1979. A Markovian queue with N servers subject to breakdown and repairs, Management Sci. 25, 849–861.
8. Vinod B., 1985. Unreliable queueing systems, Comput. Oper. Res. 12, 323–340.
9. Towsley D. and Tripathi S. K., 1991. A single server priority queue with server failures and queue flushing, Oper. Res. Lett. 10, 353–362.
10. Wartenhorst P., 1995. N parallel queueing systems with server breakdown and repair, European Journal of Operation Research 82, 302–322.
11. Wang KH. and Chang YC., 2002. Cost analysis of a finite M/M/R queueing system with balking, reneging and server breakdowns, Mathematical Methods of Operations Research 56, 169–180.
12. Haghighi AM., Medhi J., and Mohanty SG., 1986. On a multi-server Markovian queueing system with balking and reneging, Computers and Operations Research 13, 421–425.
13. Al-Seedy R.O., El-Sherbiny A.A., El-Shehawy S.A., and Ammar S.I., 2009. Transient solution of the M/M/c queue with balking and reneging, Computers & Mathematics with Applications 57, 1280–1285.
14. Plam C., 1943. Intensity fluctuations in telephone traffic, Ericsson, Technol. 1, 1–18.
15. Plam C., 1947. The distribution of repairmen in serving automatic machines, Industr. Norden, 75, 75–80.

16. Takcacs L., 1962. Introduction to the Theory of Queues, Oxford University Press, Oxford.
17. Neuts F., 1964. The distribution of the maximum length of a Poisson queue during a busy period, Oper. Res. 18, 388–395.
18. Erlander S., 1965. The remaining busy period for a single server queue with Poisson input, Oper.Res. 14, 444–459.
19. Karlin S. and McGregor J., 1958. Many server queueing processes with Poisson input and exponential service times, Pacific J. Math. 8, 87–118.
20. Rice S.O., 1962. Single server systems—II. Busy period, Bell Syst. Technol. J 41, 279–310.
21. Nance R.E., Bhat V.N., and Claybrook B.G., 1972. Busy period analysis of a time-sharing system: transform inversion, J. Assoc. Comput. Mach. 19, 453–462.
22. Shanbhag D.N., 1966. On infinite server queues with batch arrivals, J. Appl. Prob. 3, 274–279.
23. Conolly B.W., 1971. The busy period for the infinite capacity service system $M/G/\infty$. In studii diprobabilita statistica e Ricerca opertiva in onore di G. Pompilj. Instituto di calcolo delle probabilita Universita di Roma, Oderisi Gubbio, Roma, 128–130.
24. Conolly B.W., 1974. The generalised state dependent Erlangian queue: the busy period, J. Appl.Prob. 11, 618–623.
25. Bunday B.D. and EL-Badri W.K., 1985. The busy period for the M/M/1 machine interference model, Stoch. Appl. 3, 1–13.
26. Parathasarthy P.R. and Sharafali M., 1989. Transient solution to the many-server Poisson queue, J. Appl. Prob. 26, 584–594.
27. Sharma O.P., 1990. Markovian Queues, Ellis Horwood Ltd., England.
28. Kumar B.K., 1996. The busy period of an M/M/1 queue with balking, J. Appl. Stat. Sci. 3, 209–218.
29. Tarabia A. M. K., 2003. A new formula for the busy period of a non-empty multiserver queueing system, Appl. Math. Comput. 143, 401–408
30. Stadje W., 1995. The busy period of some queueing systems, Stoch. Process Appl. 55, 159–167.
31. Atencia I. and Moreno P., 2004. The discrete-time Geo/Geo/1 queue with negative customers and disasters, Computers and Operations Research 31, 1537–1548.
32. Di Crescenzo A., Giorno V., Nobile A., and Ricciardi L., 2003. On the M/M/1 queue with catastrophes and its continuous approximation, Queuing systems 43, 329–347.
33. Kumar B., Krishnamoorthy A., Madheswari S., and Basha S., 2007. Transient analysis of a single server queue with catastrophes, failures, and repairs, Queueing Systems 56, 133–141.
34. Kumar B. and Arivudainambi D., 2000. Transient solution of an M/M/1 queue with catastrophes, Computers and Mathematics with Applications 40, 1233–1240.
35. Kumar B., Vijayakumar A., and Sophia S., 2008. Transient Analysis for State-dependent Queues with Catastrophes, Stochastic Analysis and Applications 26, 1201–1217.

36. Kalidss K., Gopinath S., Gnanaraj J., and Ramanath K., 2012. Time dependent analysis of an M/M/1/N queue with catastrophes and a repairable server, Opsearch 49, 39–61.
37. Ammar Sherif I., 2014. Transient Behavior of a Two-Processor Heterogeneous System with Catastrophes, Server Failures and Repairs, Applied Mathematical Modelling 38, 2224–2234.
38. Tarabia A.M.K., 2011. Transient and steady-state analysis of an M/M/1 queue with balking, catastrophes, server failures and repairs, Journal of Industrial and Management Optimization 7, 811–823.
39. Raju SN. and Bhat UN., 1982. A computationally oriented analysis of the G / M / 1 queue, Opsearch 19, 67–83.
40. Lederman W. and Reuter GEH., 1954. Spectral theory for the differential equations of simple birth and death processes, Phil. Tran. R. Soc. London 246, 321–369.
41. Saaty T. L., 1960. Time dependent solution of the many server Poisson queue, Operations Research, 8, 755–771.

APPENDIX I

Following Raju and Bhat [39], we obtain the element of the matrix $\mathbf{B}_{k-1}^{-1}(z) = (b_{i+1,j+1}^*(z))_{(k-1) \times (k-1)}$ as:

For $i = 0, 1, 2, \ldots, k - 2,$

$$b_{i+1,j+1}^*(z) = \begin{cases} \dfrac{1}{\mu}\left[\dfrac{u_{k,j+2}(z)u_{i+1,j}(z) - u_{i+1,j+2}(z)u_{k,1}(z)}{u_{k,1}(z)} \right], & j = 0, 1, 2, \ldots, k - 3, \\[4mm] \dfrac{u_{i+1,1}(z)}{u_{k,1}(z)}, & j = k - 2, \end{cases}$$

where $u_{i,j}(z)$ are recursively given as

$$u_{i+1,i+1} = 1, \qquad j = 0, 1, 2, \ldots, k - 2;$$

$$u_{i+1,i} = \frac{z + \lambda + \mu}{\mu}, \quad i = 0, 1, 2, \ldots, k - 3;$$

$$u_{i+2,j+1} = \frac{(z + \lambda + \mu)u_{i+1,j+1} - \lambda u_{ij+1}}{\mu}, \quad j = i - 1, i - 2, \ldots, 1, 0$$

$$i = 1, 2, 3, \ldots, k - 3$$

$$u_{k,j+1} = \begin{cases} (z + \lambda + \mu)u_{k-1,j+1} - \lambda u_{k-2,j+1}, & j = i - 1, i - 2, \ldots, 1, 0, \\ z + \lambda + \mu, & j = k - 2. \end{cases}$$

APPENDIX II

If we choose Re(z)<0, sufficiently large so that, for all $i = 1, 2, \ldots, k - 1$

$$|z - z_i| \geq D_i k$$

then

$$\left| \frac{2\mu}{\omega + \sqrt{\omega^2 - r^2}}\left\{ 1 - \sum_{i=1}^{k-1} \frac{D_i}{z - z_i} \right\} \right|$$

$$= \left| \frac{\omega - \sqrt{\omega^2 - r^2}}{2\lambda_p} \right| \left| 1 - \sum_{i=1}^{k-1} \frac{D_i}{z - z_i} \right|$$

$$\leq \left| \frac{\omega - \sqrt{\omega^2 - r^2}}{2\lambda_p} \right| \left| 1 - \sum_{i=1}^{k-1} \frac{D_i}{z - z_i} \right| < 1, \text{ (see [41])}$$

Chapter 6

Studying the impact of feature importance and weighted aggregation in tackling process fairness

Guilherme Alves, Vaishnavi Bhargava, Fabien Bernier, Miguel Couceiro, and Amedeo Napoli

6.1 INTRODUCTION

Machine Learning (ML) models are increasingly present in decision support systems with critical societal impacts; for instance, in job recruitment, loan applications and criminal recidivism prediction. In spite of the objective character of these algorithmic decisions, recent studies raised fairness concerns by revealing discriminating outcomes against minorities and unprivileged groups[1,2] [1,2]. In 2016, the European Union enforced the GDPR Law[3] across all organizations and firms. The law entitles European citizens to the right to have a basic knowledge of the inner workings of ML models and their outcomes.

Two main approaches have been proposed to address algorithmic (un)fairness based on decision outcomes. One is to use fairness measures and impose fairness constraints during training [3,4], whereas the other aims to reduce the reliance of ML models on salient or sensitive features [5–7]. A natural approach to achieve the latter is to train models on datasets with these sensitive features removed. However, this may compromise the model's performance [4].

Bhargava et al. [5] proposed a human centered, model-agnostic framework that reduces classifiers' reliance on sensitive features without compromising their accuracy. LimeOut receives a triple (M, D, F) consisting of a classifier M, a dataset D and a set F of sensitive features, as input, and it outputs a classifier M' that is less reliant on the sensitive features in F. To assess the reliance of a given pre-trained model M on sensitive features, LimeOut uses a *global* variant of LIME[4] explanations [8]. If sensitive features are shown to contribute globally to M's outcomes, then M is deemed unfair. In this case, *feature dropout* is employed to build a pool of classifiers that are then aggregated to obtain an ensemble classifier.

Empirical studies [5] showed that LimeOut's ensemble models are less dependent on sensitive features when compared to original models. However, several issues concerning the use of explanation methods for assessing process fairness have been recently raised. For instance, [9,10] questioned the usefulness of explanations to assess fairness by showing that it is possible to perform

DOI: 10.1201/9781003356653-6

"adversarial attacks" to modify explanations in order to conceal unfairness issues. This led to a thorough empirical investigation [11] beyond process fairness, and where LimeOut showed consistent improvements with respect to widely used fairness metrics such as disparate impact, equal opportunity, demographic parity, equal accuracy, and predictive equality. In [11] the adaptability of LimeOut to other data types as well as to other explanation methods was also claimed. This is particularly relevant given the drawbacks of LIME explanations that have been pointed out in the literature [12,13].

In this paper, we tackle the latter issues by showing that LimeOut can be adapted to different explanation methods and aggregation functions, and by empirically observing beneficial impacts on widely used fairness metrics. More precisely, we propose FixOut, a framework that extends LimeOut by allowing any explanation method based on feature importance. To illustrate, we consider FixOut instantiated by SHAP[5] [14], an explanation method that is based on coalitional game theory, and LIME[6] to assess model fairness. Also, to construct the final ensemble model, FixOut can employ either a simple average as aggregation rule or a weighted average to take into account the global contributions of sensitive features. We thus address the three following questions:

Q1. *Are there differences between* FixOut *with LIME explanations and FixOut with SHAP explanations?*
Q2. *What is the impact of the aggregation function in the performance of* FixOut's *output models?*
Q3. *Does* FixOut *improve standard fairness metrics?*

In this chapter, we tackle each of these questions and the main contributions are the following: (1) the introduction of the FixOut framework, which is explainer-agnostic, (2) the consideration of model ensembles that take into account global contributions of sensitive features, and (3) an empirical study of FixOut on different datasets and with respect to standard fairness metrics, that illustrate the adaptability of FixOut to different explanation methods and aggregation functions.

This paper is organized as follows. After recalling the background knowledge and briefly presenting related work in Section 6.2, we describe our proposed framework FixOut in Section 6.3. Next, we present our experimental setting and discuss the results in Section 6.4. We conclude the paper in Section 6.6.

6.2 RELATED WORK

In this section, we will recall the main concepts used in this work. We first recall the notions of fairness and then we describe SHAP explanations which are used to assess fairness in FixOut.

6.2.1 Assessing fairness

The fairness of ML models can be addressed in several ways, but most fairness notions focus on models' outcomes. In this setting, there are two main approaches: one that proposes certain *fairness metrics* [3], while the other focuses on *process fairness* that assesses, for instance, the model's reliance on discriminatory or sensitive features [15], such as race, ethnicity, gender, or sexual orientation.

Fairness metrics usually rely on well-known scores measured with respect to privileged (*priv*) and unprivileged (*unp*) groups. For instance, with respect to "race", white people are usually the privileged group and vice versa. Here we recall some of the best known fairness metrics.

- **Demographic Parity (DP)** [16] is defined as the difference in the predicted positive rates between the unprivileged and privileged groups.

$$DP = P(\hat{y} = pos|D = unp) - P(\hat{y} = pos|D = priv)$$

- **Equal Opportunity (EO)** (or *disparate mistreatment*) [4] is the difference in recall scores ($\frac{TP_i}{TP_i + FN_i}$, where TP_i is true positive and FN_i is false negative for a particular group i) between the unprivileged and privileged groups.

$$EO = \frac{TP_{unp}}{TP_{unp} + FN_{unp}} - \frac{TP_{priv}}{TP_{priv} + FN_{priv}}$$

- **Predictive Equality (PE)** [17] is computed as the difference in false positive rates ($\frac{FP_i}{FP_i + TP_i}$, where FP_i is false positive for a particular group i) between unprivileged and privileged groups.

$$PE = \frac{FP_{unp}}{FP_{unp} + TP_{unp}} - \frac{FP_{priv}}{FP_{priv} + TP_{priv}}$$

- **Disparate Impact (DI)** [18] is rooted in the desire for different groups to experience similar rates of positive decision outcomes ($\hat{y} = pos$).

$$DI = \frac{P(\hat{y} = pos|D = unp)}{P(\hat{y} = pos|D = priv)}$$

- **Equal Accuracy (EA)** [16] is defined as the difference in accuracy score $\left(\frac{TP_i + TN_i}{P_i + N_i}\right)$, where TN_i is true negative of a particular group i between unprivileged and privileged groups.

$$EA = \frac{TP_{unp} + TN_{unp}}{P_{unp} + N_{unp}} - \frac{TP_{priv} + TN_{priv}}{P_{priv} + N_{priv}}$$

Inspired by the empirical setting of [9] and [5], we will focus mainly on process fairness, demographic parity, equal opportunity, and predictive equality.

6.2.2 Explanations to assess fairness

Explainability and fairness are often associated as desired qualities for trustworthy ML models [19]. In this paper, we use explanations along with fairness metrics to unveil unfairness in ML models. The explanations are obtained by explanation methods which are briefly presented in this subsection. Explanation methods differ mainly in the form of explanations or in the approach they use to generate them. For instance, Anchors provides rule-based explanations [20], while LIME [8], SHAP [14] and DeepLIFT [21] explain the outcome for a given instance by computing the contributions of a feature to the outcome. Here, we focus on model-agnostic (post-hoc) explanation methods, such as LIME and SHAP.

Given an individual prediction $f(x)$ of a data instance x by a prediction model f, explanations from both LIME and SHAP take the form of surrogate models. More precisely, they learn a linear model g – the goal of g is to locally mimic the behavior of $f(x)$ – in which each weight of g is considered as the importance of a feature for the prediction $f(x)$. The model g is then obtained by optimizing the following objective function

$$\underset{g \in \mathcal{G}}{argmin} \{L(f, g, \pi_x) + \Omega(g)\}, \tag{6.1}$$

that takes into account the complexity of explanations Ω to enforce interpretability (e.g., long explanations – i.e., those that present several features with non-zero coefficients – are not desirable) and a loss function L that is defined as

$$L(f, g, \pi_x) = \sum_{z \in Z} [f(h_x(z)) - g(z)]^2 \pi_x(z),$$

where z is the interpretable representation of x, $\pi_x(z)$ defines the interpretable neighborhood of x, and $h_x(z)$ converts z from the interpretable space to the original space. LIME and SHAP differ from the kernel employed in L and the complexity function Ω (regularizer).

6.2.2.1 LIME

Local Interpretable Model Agnostic Explanations [8] is a local post-hoc explanation method that approximates the behavior of a prediction model f w.r.t. an individual prediction $f(x)$ by fitting a linear model in the neighborhood of x. The neighborhood is obtained by perturbing the target instance x and assigning weights on the neighbors based on their proximity to x thanks to the kernel function. LIME then asks f for the corresponding predictions of the generated instances. To find a suitable linear model, LIME employs the following kernel in the objective function of Eq. 6.1

$$\pi_x(z) = exp(-\delta(x, z)^2/\sigma^2),$$

where δ is a distance function between x and interpretable instance z, and σ is the kernel width. LIME minimizes $\Omega(g)$ by reducing the number of non-zero coefficients (weights) in the linear model g.

6.2.2.2 SHAP

SHapley Additive exPlanations [14] is also a local post-hoc explanation method based on coalitional game theory. SHAP provides explanations in the form of a linear surrogate model that is defined on a "coalition" of interpretable features, and whose coefficients correspond to the contributions of the corresponding (selected) features. In the case of SHAP, these coefficients coincide with Shapley values [22]. In this paper, we focus on KernelSHAP that is a variant of SHAP. KernelSHAP receives as input an instance x, the prediction model f, and the number of coalitions m. It then learns a linear model g defined on a simplified subset of features ("coalition" that defines the representation space) by optimizing the objective function in Eq. 6.1 with the following kernel

$$\pi_x(z) = \frac{\mathcal{M} - 1}{\binom{\mathcal{M}}{|z|}|z|(\mathcal{M} - |z|)},$$

where $|z|$ is the number of present features in the coalition z and $\mathcal{M}$ is the maximum coalition size.

KernelSHAP first samples coalitions of features and it then asks for prediction of each coalition.[7] This produces a new dataset of coalitions along with predictions which is used by KernelSHAP to fit a linear model g as described in the formula:

$$g(z) = \phi_0 + \sum_{j=1}^{M} \phi_j z_j,$$

Figure 6.1 SHAP explanation for the prediction of an instance in the adult dataset.

where z_j indicates the presence/absence of the j-th feature.

To illustrate, let us consider the example of the Adult dataset where the goal is to predict if a person earns $\geq$ 50k dollars a year. Figure 6.1 presents a SHAP explanation for a prediction using Logistic Regression classifier, where the Shapley value for "Capital Gain = 2,174" is around –0.15 that indicates this feature contributes to moving the prediction toward the negative class.

6.2.3 The tension between fairness and classification performance

Mitigating unfairness raises tensions with other properties of ML models. One of them is the tension between fairness and classification performance, also known as *the fairness-accuracy trade-off* when the focus is only on classification accuracy. Taking this trade-off into consideration is crucial in many real-world problems (e.g., mortgage lending [23]) and a body of work is dedicated to studying it [24–26].

Here, instead of focusing only on well-known fairness metrics, we are interested in the tension between classification performance and process fairness. That is, our focus is also on the trade-off between classification performance and the reliance of a model on sensitive features.

Fisher et al. [27] proposed a notion of *Rashomon set*[8] as a set of ML models that present similar performances in terms of error rate (the "good models") but that utilize features differently, and they explored *model class reliance* (MCR) within a given Rashomon set. Later, Smith et al. [28] proposed an algorithmic approach to compute the MCR of Random Forest in logarithmic time. Quite interestingly, Coston et al. [29] adapted the notion of Rashomon set by integrating fairness metrics.

In this paper, we present an approach that takes into account the fairness-accuracy trade-off from the lens of process fairness. This approach is detailed next.

6.3 FIXOUT

In this section, we introduce our proposed framework FixOut and highlight the differences between FixOut and LimeOut. Similar to LimeOut [5,11], FixOut has two main components: Exp_{Global} that provides global explanations[9] in order to assess the fairness of a given pre-trained model M, and

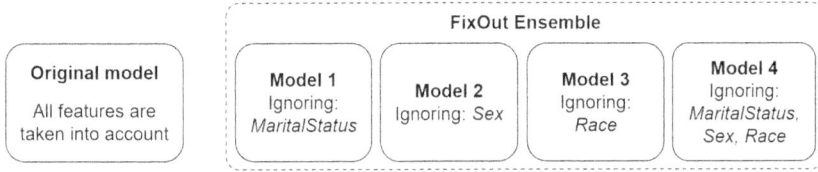

Figure 6.2 Illustration of which sensitive features are taken into account by each model.

FixOut's ensemble model that builds a fairer model M_{final} if M is deemed unfair. However, unlike LimeOut, FixOut receives as input a quadruple (M, D, F, E) where M is a pre-trained model, D is a dataset, F is a set of sensitive features, and E is an explanation method based on feature importance.

FixOut's workflow can be summarized as follows. Given (M, D, F, E), FixOut applies the component $\text{Exp}_{\text{Global}}$ using E as the explanation method. For instance, it can employ either SHAP or LIME to measure feature importance and so to evaluate the dependence of M on sensitive features (see Figure 6.2). The output is a list $F^{(k)}$ of the k most important features $a_1, a_2, ..., a_k$. As in LimeOut, FixOut applies the following rule to decide whether M is fair: if $F^{(k)}$ contains sensitive features $a_{j_1}, a_{j_2}, ..., a_{j_i}$ in F with $i > 1$, then M is deemed unfair and FixOut's second component applies; otherwise, it is considered fair and no action is taken.

In the former case (i.e., M is considered unfair), FixOut employs *feature dropout* [5] and uses the i features $a_{j_1}, a_{j_2}, ..., a_{j_i} \in F$ to build a pool of $i + 1$ classifiers in the following way: for each $1 \le t \le i$, FixOut trains a classifier M_t after removing a_{j_t} from D, and an additional classifier M_{i+1} trained after removing all sensitive features F from D. As in LimeOut, this pool of classifiers is used to construct an ensemble classifier M_{final} (see Figure 6.2). However, instead of a simple average, FixOut employs a weighted average using weights that take into account a feature's contributions. Let $c'_{j_t} \in [0,1]$ be the normalized global feature contribution associated with a_{j_t}. We standardize feature contributions by $c'_{j_t} = \frac{a_{j_t} - min(F^{(k)})}{max(F^{(k)}) - min(F^{(k)})}$, where $min(F^{(k)})$ and $max(F^{(k)})$ are the lowest and the highest feature contributions among $F^{(k)}$, respectively. Now, let us define the weights w_t of M_t and the weight w_{i+1} of M_{i+1} as

$$ w_t = \frac{c'_{j_t}}{1 + \sum_{u=1}^{i} c'_{j_u}}, \quad 1 \le t \le i, \quad \text{and} \quad w_{i+1} = \frac{1}{1 + \sum_{u=1}^{i} c'_{j_u}}. $$

The main idea behind using feature contribution in the weighted averaging rule is to ensure higher weights for classifiers trained without sensitive features whose contributions to M's outcomes are high. Also, the additional classifier M_{i+1}, the one that is trained without any sensitive feature, receives high weight. For a data instance x and a class C, the ensemble classifier M_{final} uses the following rule to predict the probability of x being in class C,

$$P_{M_{final}}(x \in C) = \sum_{t=1}^{i+1} w_t P_{M_t}(x \in C), \tag{6.2}$$

where $P_{M_t}(x \in C)$ is the probability predicted by model M_t.

Example: To illustrate how FixOut works, let us consider the example taken from an experiment on the Adult dataset using AdaBoost.[10] The dataset contains more than 32,000 instances; each instance is described by 14 features. The goal is to predict if a person earns ≥ 50k dollars a year.

We consider as sensitive features: "Marital Status", "Sex" and "Race". Figure 6.2 presents how the ensemble model is constituted and which features are dropped. Note that each classifier in the ensemble is fairer than the original model, since each one drops at least one sensitive feature while the original model takes all sensitive features into account.

Table 6.1 presents the lists of the $k = 10$ most important features with their respective global contributions of both original and ensemble models using SHAP explanations (F+SHAP and Fw+SHAP). One notes that FixOut's ensemble classifiers were less reliant on sensitive features. For instance, in the experiment using AdaBoost on the Adult dataset, the absolute value of "MaritalStatus" decreased from −2.439179 (Original +SHAP) to −1.156453 (F+SHAP) and −0.200202 (Fw+SHAP).

6.4 EXPERIMENTS

In this section, we briefly present the datasets and the experimental setting that we used to perform our experiments. We then examine the obtained results in the following way. First, we report and compare the obtained classification measures on several classifiers and on FixOut's ensembles in Subsection 6.4.2. We then assess process fairness using LIME and SHAP explanations in Subsection 6.4.3. Finally, we evaluate fairness using standard metrics in Subsection 6.4.4.

6.4.1 Datasets and experimental setup

The experiments[11] were conducted on three datasets used in [11], namely, German,[12] Adult,[13] and LSAC.[14] All datasets share common characteristics that allow us to run our experiments: a binary class label and the presence of sensitive features.[15] Table 6.2 summarizes basic information about these datasets.

In order to perform our experiments, we made 50 different train/test splits according to the following rule; for each experiment, we used split

Table 6.1 Global explanation of AdaBoost on adult dataset

Original (SHAP)		F+SHAP		Fw+SHAP	
Feature	Contrib.	Feature	Contrib.	Feature	Contrib.
MaritalStatus	−2.439179	Education	1.979514	Education	1.930672
Education	1.892603	Age	1.698045	Age	1.734298
Age	1.754097	**MaritalStatus**	−1.156453	Relationship	−1.407386
Occupation	−1.610042	CapitalGain	−0.572854	Occupation	−0.799977
Relationship	1.114786	Hoursperweek	0.399115	CapitalGain	−0.559822
Education-Num	−0.848503	Occupation	−0.392517	Education-Num	0.532497
CapitalGain	−0.583551	**Sex**	0.304267	Hoursperweek	0.409731
Sex	0.436266	Relationship	−0.294556	**Sex**	0.254732
Hoursperweek	0.382647	Education-Num	0.235512	**MaritalStatus**	−0.200202
Race	0.173112	CapitalLoss	−0.131449	Workclass	−0.148899

Table 6.2 Datasets employed in the experiments

Dataset	# features	# instances	Sensitive features
German	20	1000	"statussex", "telephone", "foreign worker"
Adult	14	32561	"MaritalStatus", "Race", "Sex"
LSAC	11	26551	"race", "sex", "family_income"

data into 70% training set and 30% testing. As the datasets are imbalanced, we used Synthetic Minority Oversampling Technique (SMOTE[16]) over training data to generate the samples synthetically. We trained original and FixOut's ensemble models on the balanced (augmented) datasets using five classifiers. We used Scikit-learn implementation of the following algorithms: AdaBoost(**ADA**), Bagging (**BAG**), Random Forest (**RF**), and Logistic Regression (**LR**). We kept the default parameters of Scikit-learn documentation. In order to estimate Shapley values faster, especially in the presence of continuous features, we used K-means clustering with the number of clusters $n = 10$ to reduce feature domains; otherwise, the full domain is considered.

Driven by the questions presented in Section 6.1, we defined four instantiations of FixOut, namely: F+SHAP, Fw+SHAP, F+LIME, and Fw +LIME. F+SHAP uses SHAP as explanation method and simple average as aggregation function. Fw+SHAP also employs SHAP but instead of applying the simple average rule, it employs the weighted average function described in Section 6.3. Similar to F+SHAP, F+LIME uses the simple average but with LIME explanations. The latter, Fw+LIME, uses LIME as explainer but with the weighted average function. We also define baseline model as the classifier trained without any sensitive feature.

6.4.2 Classification performance assessment

Table 6.3 shows the classification assessment obtained throughout the performance of the experiments. For each dataset, we have the average precision and recall of original models and of FixOut's ensemble models. In this analysis, we consider the following instantiations of FixOut: F+SHAP, Fw+SHAP and Fw+LIME. We did not look at F+LIME as it has equivalent results to F+SHAP.

We performed the Kruskal-Wallis test (with 95% of confidence) to assess whether the results were statistically significant among the average precision and recall of original and FixOut's ensemble models (F+SHAP, Fw +SHAP, and Fw+LIME).[17] We found that the experiments on LSAC dataset were statistically significant, w.r.t all measures and using all classifiers. However, we did not find the same on the German dataset, where only the experiments using BAG showed statistically significant results. In the case of

Table 6.3 Classification assessment. Highlighted cells indicate statistical significance according to Kruskal-Wallis test with 95% of confidence

Dataset	Method	Precision				Recall			
		ADA	BAG	LR	RF	ADA	BAG	LR	RF
German	Baseline	.5818	.5436	.5776	.6775	.4815	.5545	.5523	.3463
	Original	.5707	.5124	.5716	.6883	.5317	.5738	.5495	.3595
	F+SHAP	.5801	.5549	.5754	.7060	.5321	.5371	.5622	.3585
	Fw+SHAP	.5809	.5537	.5746	.7003	.5390	.5142	.5632	.3708
	Fw+LIME	.5764	.5471	.5708	.7019	.5373	.5076	.5602	.3541
Adult	Baseline	.6888	.6407	.4011	.6943	.6902	.6749	.5269	.6182
	Original	.6884	.6419	.3857	.7004	.6882	.6687	.5600	.6175
	F+SHAP	.6930	.6838	.3859	.7121	.6856	.6451	.5317	.6153
	Fw+SHAP	.6948	.6805	.4213	.7104	.6829	.6343	.5123	.6119
	Fw+LIME	.6927	.6771	.3922	.7108	.6838	.6343	.5398	.6156
LSAC	Baseline	.9097	.8881	.8658	.8865	.8483	.9247	.7596	.9243
	Original	.8986	.8846	.8548	.8771	.9016	.9413	.8330	.9473
	F+SHAP	.9044	.8898	.8596	.8838	.8850	.9448	.8136	.9436
	Fw+SHAP	.9080	.8894	.8674	.8866	.8599	.9433	.7584	.9280
	Fw+LIME	.9071	.8909	.8614	.8851	.8618	.9201	.7899	.9363

the Adult dataset, we found that the experiments using ADA, BAG, and RF were statistically significant.

Our analysis is now based on the comparison between the classification measures of the original and ensemble models. We notice that FixOut's ensemble models improve, or at least maintain, the precision among the results that were statistically significant. However, we also observe that the recall decreased in the same experiments. On those that were not were statistically significant, the results indicate that FixOut's ensemble models maintain the classification measures even though we can see a slight improvement from the original model to one of FixOut's ensemble models.

6.4.3 Process fairness assessment

We now address process fairness, namely, the reliance of FixOut's ensemble outputs on sensitive features. To demonstrate the ability of FixOut to reduce the reliance of classifiers on sensitive features, regardless of the choice of explanation method, we performed several experiments using LIME and SHAP explanations. The idea was to verify that by changing the explanation method (from LIME to SHAP) we can still demonstrate the impact of feature dropout on the model's reliance on sensitive features. This empirical analysis helps us to answer Q1 and Q2.

Figures 6.3a, 6.3b and 6.3c show the frequency of sensitive features in the top ten most important features for both pre-trained classifiers and FixOut's ensemble models (F+SHAP, F+LIME, Fw+SHAP, Fw+LIME). We can observe that FixOut decreased the frequency of sensitive features in the ensemble models. We compare the original models (pre-trained classifiers with explanations obtained from LIME and SHAP, namely Original+LIME and Original+SHAP, respectively) with ensemble models produced by FixOut using different explanation methods (F+SHAP and F+LIME) and different aggregation functions (Fw+SHAP and Fw+LIME). We notice that, on the German dataset, FixOut drastically reduced the

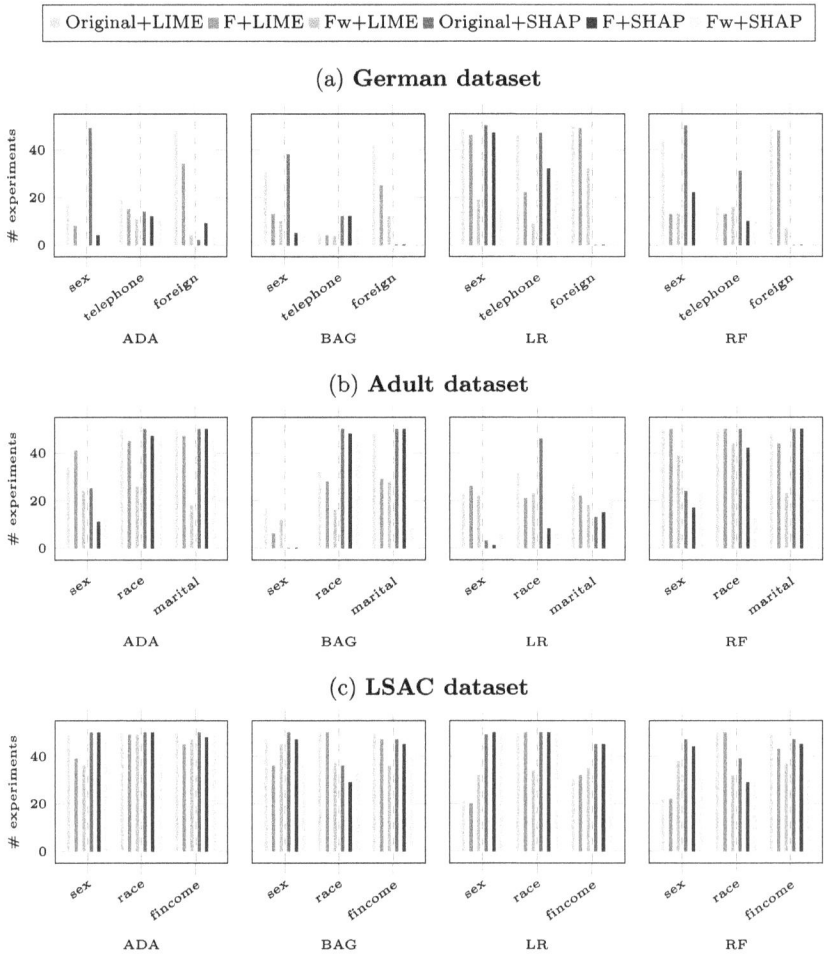

Figure 6.3 Frequency of sensitive features in the top ten most important features for different datasets. The dashed lines separate LIME and SHAP explanations for each sensitive feature.

frequency of sensitive features in the top ten. One exception is for the sensitive feature "foreign" using AdaBoost.

We now analyze the contribution of sensitive features rather than look at the frequency of them in the top ten most important features. Table 6.4 shows the average contribution of each sensitive features throughout 50 experiments. Again, we can observe that FixOut reduces the dependency of models on sensitive features since the average contribution of the sensitive features decreased from the original model to FixOut's ensemble models. We can also observe that it happened independently of the choice of explanation method. In other words, we noticed the same behavior when either LIME or SHAP is applied. Also, the use of a weighted aggregation rather than a simple aggregation (e.g., Fw+SHAP instead of F+SHAP) reduces the dependence of the model on sensitive features even more.

6.4.4 Fairness metrics assessment

In this section, we assess fairness using the standard metrics introduced in Subsection 6.2.1 in order to have a different perspective of the fairness of FixOut's ensemble models. We computed Demographic Parity (DP) and Equal Opportunity (EO) using IBM AI Fairness 360 Toolkit[18] [30]. We also considered Predictive Equality (PE) [17] to measure the false positive differences between privileged and unprivileged groups. The metrics DP, EO, and PE give values in the interval [−1,1] where 0 indicates a perfect fair model. This empirical analysis is linked with the third question (Q3).

Fairness metrics are depicted in Figures 6.4, 6.5, and 6.6. In this analysis, we compare the original and FixOut's ensemble models based on fairness metrics for each combination of classifier and sensitive feature.

Violet triangle points indicate the values for original models, while magenta points, purple points and blue points represent the values for Fw +LIME, F+SHAP and Fw+SHAP, respectively. The dashed line is the reference for a fair model (optimal value). In this analysis, we do not compare the results with Original+LIME and F+LIME as both are equivalent to Original+SHAP and F+SHAP, respectively, regarding the fairness metrics. In other words, the pairs (Original+SHAP, Original +LIME) and (F+SHAP, F+LIME) have the same fairness metric values because explanation methods do not affect the computation of fairness metrics. In these cases, the aggregation function does not depend on the feature contributions obtained from explanations. We then simplified this analysis by ignoring Original+LIME and F+LIME.

Results for the German dataset are depicted in Figure 6.4. Like in LimeOut experiments, FixOut produces ensemble models that are fairer according to metrics DP and EO, since magenta, purple and blue points are closer to zero compared to the triangle violet points (pre-trained model). However, unlike in LimeOut's case, we can not see the same behavior for ensemble models compared to original models according to PE metric.

Table 6.4 Average contribution of sensitive features

Dataset	Method	ADA			BAG			LR			RF		
German		sex	telephone	foreign	sex	telephone	foreign	sex	telephone	foreign	sex	telephone	foreign
	Original+LIME	-0.13	0.12	3.84	-2.13	0.33	6.36	-13.90	10.08	25.55	-3.29	0.85	23.00
	F+LIME	-0.05	0.09	0.85	-0.63	0.15	1.88	-7.46	2.86	11.90	-0.55	0.67	7.47
	Fw+LIME	0.00	0.06	0.02	-0.79	0.11	0.65	-2.00	1.24	3.28	-0.49	0.69	0.23
	Original+SHAP	-0.68	0.10	0.01	-5.13	1.55	0.00	-31.20	11.59	0.00	-10.53	3.21	0.00
	F+SHAP	-0.02	0.08	0.04	-0.76	1.08	0.00	-10.20	3.52	0.00	-1.87	0.69	0.00
	Fw+SHAP	-0.07	0.08	0.13	-0.87	0.71	0.00	-1.37	3.25	0.06	-1.87	0.69	0.00
Adult		marital	race	sex	marital	race	sex	marital	race	sex	marital	race	sex
	Original+LIME	0.88	0.43	0.48	14.35	-1.02	2.11	0.49	-0.05	1.13	8.10	11.88	9.59
	F+LIME	0.34	0.25	0.21	2.90	-0.40	3.03	-0.01	-0.11	0.07	4.63	5.77	5.80
	Fw+LIME	-0.02	0.13	0.03	-0.65	-0.62	2.57	0.01	-0.07	0.34	1.05	1.45	1.70
	Original+SHAP	-3.32	0.08	0.53	98.35	0.00	4.10	-0.15	0.00	1.67	-23.29	1.29	9.25
	F+SHAP	-1.26	0.03	0.24	31.51	0.00	5.25	-0.06	0.00	0.06	-11.68	0.86	3.13
	Fw+SHAP	-0.17	0.03	0.18	0.37	0.00	4.47	-0.21	-0.02	0.02	-1.73	1.30	0.55
LSAC		sex	race	fincome	sex	race	fincome	sex	race	fincome	sex	race	fincome
	Original+LIME	0.01	0.38	0.33	2.38	-28.79	13.56	-1.35	25.30	-5.01	-0.31	-43.64	-3.48
	F+LIME	0.02	0.18	0.15	1.18	-18.02	6.31	-0.59	7.89	-1.08	-0.28	-15.18	-1.18
	Fw+LIME	0.02	0.09	0.06	1.72	-0.29	1.81	-0.57	0.95	0.09	-0.27	0.38	-0.65
	Original+SHAP	-0.06	0.05	-0.07	-6.59	-0.58	-2.43	-8.63	5.38	1.28	-3.31	-0.18	-2.70
	F+SHAP	-0.06	0.02	-0.05	-2.44	-0.82	-1.58	-3.50	2.74	0.41	-1.71	-0.67	-1.65
	Fw+SHAP	-0.02	0.01	-0.01	-0.42	-0.91	-0.49	-1.03	1.17	0.29	-0.26	-0.67	-0.43

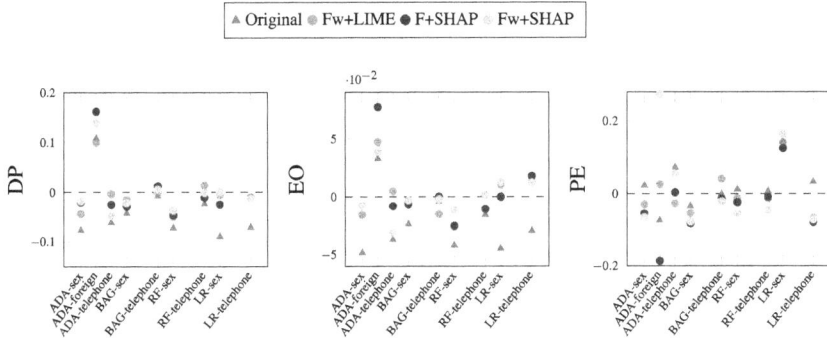

Figure 6.4 Fairness metrics for German Dataset.

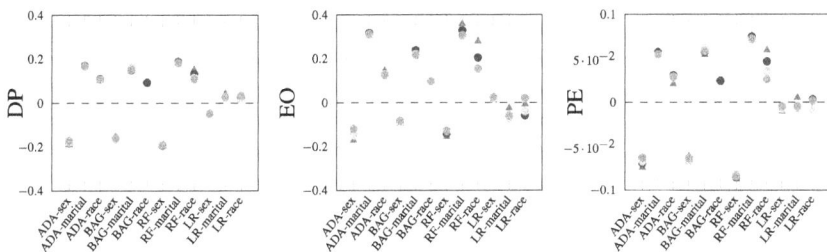

Figure 6.5 Fairness metrics for Adult Dataset.

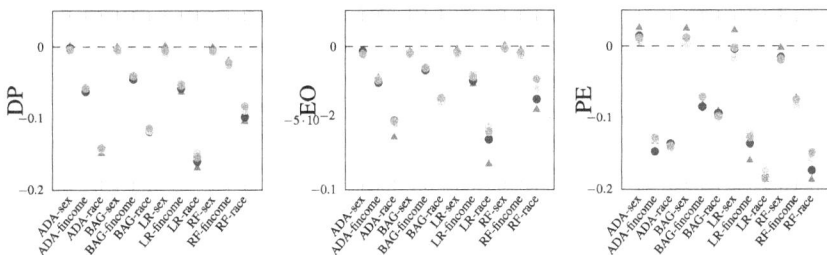

Figure 6.6 Fairness metrics for LSAC Dataset.

Figure 6.5 shows the results on fairness metrics for the Adult dataset. In this dataset, FixOut ensemble models keep values of all metrics in almost scenarios. We only see a deterioration of fairness when we computed EO for Logistic Regression focusing on "marital status" and also on "race". Results depicted in Figure 6.6 concern the LSAC dataset. Again, FixOut's ensemble models keep values of all metrics in almost scenarios, except in the case of PE for Random Forest on "sex". This behavior means that FixOut at least maintains the value of fairness metrics when it reduces the dependence on sensitive features, but it cannot ensure fairness metrics closer to the optimal value in all cases.

We also performed the Kruskal-Wallis test (with 95% of confidence) to assess whether the obtained results were statistically significant among the four different methods (Original+SHAP, Fw+LIME, F+SHAP, Fw+SHAP) for each combination classifier and sensitive feature. We noticed that some sensitive features are more impacted than others. For instance, the statistical test showed that the difference of fairness metrics computed on the sensitive feature "sex" were statistically significant.

It also showed that FixOut is able to change the bias for this particular sensitive feature. However, we did not observe the same behavior for other sensitive features. According to the same statistical test, FixOut had a significant impact when used on Random Forest models than other models.

These results indicated that FixOut improves, or at least maintains, the fairness metrics DP and EO for ensemble models. However, it also showed that there is still space for improvements. The aggregation rule should be learnt to further improve FixOut's ensemble models with respect to standard fairness metrics.

6.5 EXTENSIONS OF FIXOUT

In this section, we briefly point out some extensions and improvements that have been made on FixOut. We start by an approach to discover the value for the parameter k and then we propose an extension of FixOut to text data.

6.5.1 Automating the choice of k: the algorithm Find-K

FixOut users must know beforehand a suitable value for k so that FixOut builds L with the k most important features. This list is used in the fairness assessment; the framework only takes into account the first k most important features. An inappropriate value for k prevents FixOut from correctly detecting unfairness issues as all sensitive features may not appear in L. We have proposed an algorithm that automates the choice of k, i.e FixOut [31]. The algorithm is based on the statistical measure *kurtosis*, which indicates the flatness of a distribution.

Find-K selects a sub-sample of features of L in the following way. It iteratively removes features from L (from the less important to the most important) by analyzing the kurtosis $\gamma(L)$ before and after the deletion of a subset of features. Once a feature is removed, the kurtosis of L changes. The algorithm stops when $|\gamma(L) - \gamma(L')| > \alpha$, where L is the original list, L' is the list obtained after removing features, and $\alpha > 0$ is a parameter.

6.5.2 FixOut's extension for textual data

FixOut has been also extended to textual data and it has been employed in the task of classifying tweets as hate speech or not [31]. As language variant should not be a criteria for classifying a tweet as hate speech or not, FixOut was extended in order to reduce the dependence of classifiers on certain words. To do so, feature dropout was adapted to word dropout. However, once the number of words to be removed increases, FixOut becomes less effective. In order to overcome this issue, instead of ignoring a single word, words are grouped in order to do a "bag of word dropout", i.e., the classifier drops several words. The contribution of words using bag of words dropout (grouping words) was lower than word dropout (without grouping words) [31].

6.6 DISCUSSION AND CONCLUSION

We have proposed FixOut, a human-centered and model-agnostic framework to make ML models fairer. FixOut was proposed to address and tackle process fairness: it first assesses the dependence of a given pre-trained ML model on sensitive features by global explanations in the form of feature contributions to the model's outcomes. If the model is shown to rely on sensitive features, then FixOut employs feature dropout followed by an ensemble approach to produce a new model.

The empirical study was driven by the three questions stated in Section 6.1, and it was performed on real datasets. It showed that FixOut drastically reduces a model's reliance on sensitive features (i.e., improved process fairness) without compromising accuracy and precision, regardless of the choice of explanation method. It also showed a significant decrease in both the feature contributions and the frequencies of sensitive features in the top ten of most important features when using SHAP or LIME explanations. In fact, we observed a similar behavior between FixOut with SHAP and FixOut with LIME with respect to classification performances and process fairness (Q1). We also evaluated the impact of two different aggregation functions on FixOut's output models. The comparison of the aggregation rules employed, namely, the simple average and the weighted average rules, and revealed that the latter rule had a beneficial impact on process fairness (Q2). Even though FixOut was designed to improve process fairness, we also assessed FixOut empirically using well-known fairness metrics. Our results showed overall improvements with respect to some fairness metrics and sensitive features. However, FixOut did not guarantee improvements in all cases (Q3) and this asks for a deeper study in this direction.

ACKNOWLEDGMENTS

This research was partially supported by TAILOR, a project funded by EU Horizon 2020 research and innovation programme under GA No 952215, and the Inria Project Lab "Hybrid Approaches for Interpretable AI" (HyAIAI).

NOTES

1 https://www.reuters.com/article/us-amazon-com-jobs-automation-insight-idUSKCN1MK08G
2 https://www.bbc.com/news/business-50365609
3 General Data Protection Regulation: https://gdpr-info.eu/
4 LIME stands for Local Interpretable Model-agnostic Explanations.
5 SHAP stands for SHapley Additive exPlanations.
6 The instantiation of FixOut with LIME explanations and the simple average aggregation is equivalent to LimeOut.
7 Before asking for predictions, KernelSHAP converts a coalition z from the representation (interpretable) space to the original space using $h_x(z)$.
8 Breiman [7] used "Rashomon effect" to multiple functions with similar error rates but different descriptions.
9 Essentially, we use sampling techniques to choose representative data instances: LimeOut uses submodular pick [23], whereas FixOut uses a simple bootstrapping approach available in the implementation of SHAP. These are then aggregated to obtain a global ranking of feature contributions.
10 http://archive.ics.uci.edu/ml/datasets/Adult
11 For a demonstration of the selected datasets, please visit: https://fixout-app.loria.fr.
12 The goal is to predict if an applicant has a high credit risk. Available at https://archive.ics.uci.edu/ml/datasets/statlog+(german+credit+data)
13 The goal is to predict whether a person earns more than 50k dollars per year. Available at http://archive.ics.uci.edu/ml/datasets/Adult
14 The goal is to predict whether a law student passes in the bar exam. Available at http://www.seaphe.org/databases.php
15 For each dataset, we performed a correlation analysis among pairs of features. The goal was to identify if there exist sensitive features highly correlated with non-sensitive features. We employed the Pearson correlation coefficient and we did not find any pair (sensitive, non-sensitive) of features that has a correlation ≥ 0.75 in all datasets considered.
16 https://machinelearningmastery.com/threshold-moving-for-imbalanced\ \-classification/
17 F+SHAP and F+LIME present the same classification performance since both use the simple average rule. In order to simplify this part of our analysis we ignore F+LIME
18 https://github.com/Trusted-AI/AIF360

REFERENCES

1. Angwin, J., et al.: Machine bias: There's software used across the country to predict future criminals and its biased against blacks. ProPublica (2016).
2. Guegan, D., et al.: Credit risk analysis using machine and deep learning models. In: Risks. vol. 6, p. 38 (2018).
3. Speicher, T., et al.: A unified approach to quantifying algorithmic unfairness: Measuring individual &group unfairness via inequality indices. In: ACM SIGKDD. pp. 2239–2248 (2018).

4. Zafar, M. B., et al.: Fairness constraints: Mechanisms for fair classification. In: Artificial Intelligence and Statistics. pp. 962–970 (2017).
5. Bhargava, V., et al.: LimeOut: An ensemble approach to improve process fairness. In: ECML PKDD Int. Workshop XKDD. pp. 475–491 (2020).
6. Breiman, L.: Statistical modeling: The two cultures. Statistical Science 16(3), 199–231 (2001).
7. Grgic-Hlaca, N., et al.: The case for process fairness in learning: Feature selection for fair decision making. In: NIPS Symposium on Machine Learning and the Law. p. 2 (2016).
8. Ribeiro, M. T., et al.: "Why Should I Trust You?": Explaining the predictions of any classifier. In: ACM SIGKDD. pp. 1135–1144 (2016).
9. Dimanov, B., et al.: You shouldn't trust me: Learning models which conceal unfairness from multiple explanation methods. In: ECAI. pp. 2473–2480 (2020).
10. Slack, D., et al.: Fooling lime and shap: Adversarial attacks on post hoc explanation methods. In: AAAI/ACM AIES. pp. 180–186 (2020).
11. Alves, G., et al.: Making ML models fairer through explanations: the case of LimeOut. In: AIST'20. pp. 3–18 (2020).
12. Garreau, D., Luxburg, U.: Explaining the explainer: A first theoretical analysis of LIME. In: AISTATS. pp. 1287–1296 (2020).
13. Molnar, C.: Interpretable machine learning: A guide for making black box models explainable. https://christophm.github.io/interpretable-ml-book/ (2018).
14. Lundberg, S. M., Lee, S.: A unified approach to interpreting model predictions. In: NIPS. pp. 4765–4774 (2017).
15. Grgić-Hlača, N., et al.: Beyond distributive fairness in algorithmic decision making: Feature selection for procedurally fair learning. In: AAAI. pp. 51–60 (2018).
16. Hardt, M., et al.: Equality of opportunity in supervised learning. In: NIPS. pp. 3323–3331 (2016).
17. Makhlouf, K., et al.: On the applicability of ml fairness notions. arXiv preprint arXiv:2006.16745 (2020).
18. Dwork, C., et al.: Fairness through awareness. In: Proceedings of the 3rd Innovations in Theoretical Computer Science Conference. pp. 214–226 (2012).
19. Guidotti, R., et al.: A survey of methods for explaining black box models. ACM computing surveys (CSUR) 51(5), 1–42 (2018).
20. Ribeiro, M. T., et al.: Anchors: High-precision model-agnostic explanations. In: AAAI. pp. 1527–1535 (2018).
21. Shrikumar, A., et al.: Learning important features through propagating activation differences. ICML pp. 3145–3153 (2017).
22. Shapley, L. S.: A value for n-person games. In: Contributions to the Theory of Games. pp. 307–317 (1953).
23. Lee, M. S. A., Floridi, L.: Algorithmic fairness in mortgage lending: From absolute conditions to relational trade-offs. Minds and Machines 31(1), 165–191 (2021).
24. Bechavod, Y., Ligett, K.: Learning fair classifiers: A regularization-inspired approach. CoRR abs/1707.00044 (2017), http://arxiv.org/abs/1707.00044
25. Menon, A. K., Williamson, R. C.: The cost of fairness in binary classification. In: Conference on Fairness, Accountability and Transparency. pp. 107–118. PMLR (2018).

26. Zliobaite, I.: On the relation between accuracy and fairness in binary classification. arXiv preprint arXiv:1505.05723 (2015).
27. Fisher, A., et al.: All models are wrong, but many are useful: Learning a variable's importance by studying an entire class of prediction models simultaneously. JMLR **20**(177), 1–81 (2019).
28. Smith, G., et al.: Model class reliance for random forests. NeurIPS **33**, 22305–22315 (2020).
29. Coston, A., et al.: Characterizing fairness over the set of good models under selective labels. arXiv preprint arXiv:2101.00352 (2021).
30. Bellamy, R. K., et al.: Ai fairness 360: An extensible toolkit for detecting and mitigating algorithmic bias. IBM Journal of Research and Development **63**, 4-1 (2019).
31. Alves, G., et al.: Reducing unintended bias of ml models on tabular and textual data. The 8th IEEE International Conference on Data Science and Advanced Analytics (2021).

Chapter 7

Gaussian mixture model with modified hard EM algorithm in clustering problems

Samyajoy Pal and Christian Heumann

7.1 INTRODUCTION

In many modern-day statistical and machine learning problems the expectation maximization (EM) algorithm [1] is widely used. Applications of the EM algorithm in unsupervised learning often involve mixture models [2,3]. Use of mixture models can be seen in many fields such as image matching [4] and audio and video scene analysis [5]. But there exists an alternative approach for estimating parameters of mixture models, which is known as Hard EM. Hard EM or Viterbi Training (VT) was first introduced by [6] for speech recognition technique. Since then, many researchers have used Hard EM for speech recognition problems [7–11]. In other fields such as natural language processing (NLP) [12–18], bioinformatics [19–21] and image analysis [22], it is also extensively used.

Hard EM is an unsupervised learning technique, which can be seen as a coordinate ascent procedure that locally optimizes a function. In the case of mixture models, Hard EM is often described as Classification EM (CEM) as CEM maximizes the classification likelihood instead of the mixture likelihood [23,24]. Neal, Hinton [25] has called another version of Hard EM as Sparse EM where, in the E step, the algorithm, instead of finding the marginals like standard EM, finds the modes of the hidden variables. Hard EM is also linked with KMeans [26] clustering algorithm. KMeans can be seen as a special case of Hard EM for a mixture of Gaussians with a common covariance matrix of the form $\sigma^2 I$ and unknown σ [24,27].

In many situations, the EM algorithm becomes slow and computationally expensive. On the other hand, Hard EM provides an easy and computationally less intensive solution by an appropriate maximization step [28]. It is also known for being more robust and faster than standard EM [29]. Despite having all these desired qualities, Hard EM has some theoretical disadvantages for which it is assumed to be less accurate than standard EM [30]. Contrary to standard EM, Hard EM does not increase the likelihood of the parameters given the observed data x. Instead, it increases the joint likelihood of latent variables and parameters. And that is why it lacks

DOI: 10.1201/9781003356653-7

consistency [31] and, in fact, can produce biased estimates [32]. Even with the above drawbacks, Hard EM still enjoys a fair share of applications in practice [33]. However, when and under what circumstances, Hard EM should be preferred over standard EM remains an open problem even today [29], which calls for further investigation. Another issue with EM algorithm is that sometimes it converges to local optimum instead of a global one [34]. To overcome the issue, generally two techniques are used. One involves using repeated random initialization and another uses KMeans centroids and empirical standard Deviations as starting values [35]. However, repeated run of EMs with random initialization consumes more time and in many situations (e.g., imbalanced data sets) KMeans work poorly. As a result, using values obtained from poorly fitted models lead to poor performance of EM algorithm.

In our study, we have revisited the problem of Hard EM in the case of mixture models, more precisely for its applications in clustering. Despite having theoretical disadvantages over standard EM, we wanted to investigate if Hard EM really performs worse than standard EM for clustering problems in different situations. The main objectives of our study are to:

- provide a modification to Hard EM to stop fast convergence to local optimums with one or more clusters being empty.
- assess the performance of Hard EM in clustering for different situations (e.g., increasing number of clusters, increasing dimensions, increasing overlap of clusters, imbalance in data points, etc.)
- compare the performance of Hard EM with standard EM to investigate if it really works worse as assumed.

We have used Hard EM with some modifications to build a Gaussian mixture model (Hard GMM). The model has been used on five benchmark data sets for Gaussian mixtures [36] which are often preferred to test novel clustering methods. We have evaluated the performance of the model in different situations and compared it with the standard EM (Usual GMM) at each stage. We have also used two real data sets from biology to evaluate its performance.

7.2 METHODOLOGY

In this section, we would like to introduce the mixture model in general using Hard EM.

Let $X_1, X_2 \ldots , X_N$ denote a random sample of size N, where X_i is a p dimensional random vector with probability density function $f(x_i)$ on R^p. We can write $X = (X_1^T, ..., X_N^T)^T$, where the superscript T denotes vector transpose.

Note that, the entire sample is represented by X, i.e., X is a N – tuple of points in $\mathbb{R}^p$ or an $N \times p$-matrix. $X = (x_1^T, ..., x_N^T)^T$ denotes an observed random sample where x_i is the observed value of the random vector X_i.

The density of a mixture model with k components for one observation x_i is given by the mixture density

$$p(x_i) = \sum_{j=1}^{k} \pi_j f_j(x_i | \alpha_j) \tag{7.1}$$

where $\pi = (\pi_1, ... , \pi_k)$ contains the corresponding mixture proportions with $\sum_{i=1}^{k} \pi_i = 1$ and $0 \leq \pi_i \leq 1$. $f_j(x_i | \alpha_j)$ is the density component of mixture j and α_j, $j = 1, 2, ... , k$, are vectors of component specific parameters for each density. Then $\alpha = (\alpha_1, ... , \alpha_k)$ denotes the vector of all parameters (except π) of the model. The log likelihood of the model for a sample of size N is then given by

$$\log p(x_1, ..., x_N | \alpha, \pi) = \sum_{i=1}^{N} \log \left[\sum_{j=1}^{k} \pi_j f_j(x_i | \alpha_j) \right]. \tag{7.2}$$

The parameters can be estimated using the EM algorithm with some modifications. For that purpose, let us introduce latent variables Z_i, which are categorical variables taking on values $1, ... , k$ with probabilities $\pi_1, ... , \pi_k$ such that $Pr(X_i | Z_i = j) = f_j(x_i)$, $j = 1, ... , k$.

Further, probabilities γ_{ij} are introduced (conditional on the observed data $X = x$ and the parameter α):

$$\gamma_{ij}(x_i) = Pr(Z_i = j | X = x, \alpha) = \frac{\pi_j f_j(x_i | \alpha_j)}{\sum_{j=1}^{k} \pi_j f_j(x_i | \alpha_j)}. \tag{7.3}$$

Equation 7.3 can be seen as the probability of cluster membership j for a data point x_i. Now, we must note that, Hard EM and standard EM optimize two different objective functions. In case of a Hard EM the following objective function is optimized.

$$\hat{\Theta} = \underset{\Theta}{\operatorname{argmax}} \max_{z_1, ..., z_N} P_{\Theta}(x_1, ..., x_N, z_1, ..., z_n) \tag{7.4}$$

where, Θ denotes all parameters (π, α). But in case of a standard EM, the objective function is

$$\hat{\Theta} = \underset{\Theta}{\operatorname{argmax}} \sum_{z_1, ..., z_N} P_{\Theta}(x_1, ..., x_N, z_1, ..., z_n) \tag{7.5}$$

Hard EM does not maximize the likelihood; instead it applies a delta function approximation to the posterior probabilities $Pr(Z_i = j | X = x, \alpha)$, where $Z_i, i = 1, \dots, N$ are the latent variables representing class labels. X and α are the data and model parameters, respectively. The approximation changes the E step as follows,

$$Pr\,(Z_i = j | X = x, \alpha) \approx I\,(j = z_i^*)\qquad(7.6)$$

where, $z_i^* = \text{argmax }\gamma_{ij}$. γ_{ij}'s are nothing but the responsibilities (probabilities) for j, each data point belonging to different clusters. After this step, for a standard hard EM, the ratio of empirical cluster members and total observations serve as the new estimates of π_j. However, for our modified hard EM we do not propose to use that estimate. Instead, we optimize the expected complete data log likelihood with respect to π_j as usual like a standard EM. Our proposed technique to obtain the estimates of α_j is explained below.

The expectation of complete data log likelihood is given by

$$Q\,(\alpha, \alpha^{t-1}) = E\left[\sum_{i=1}^{N} \log(p\,(x_i, z_i | \alpha)) | x, \alpha^{t-1}\right],\qquad(7.7)$$

where t is the current iteration number. It can also be shown that [37]

$$Q\,(\alpha, \alpha^{t-1}) = \sum_{i=1}^{N}\sum_{j=1}^{k} \gamma_{ij} \log \pi_j + \sum_{i=1}^{N}\sum_{j=1}^{k} \gamma_{ij} \log f_j\big(x_i | \alpha_j\big)\qquad(7.8)$$

At the M step, we optimize Q with respect to π and α. π_j is estimated in the usual way by $\frac{N_j}{N}$, where, $N_j = \sum_{i=1}^{N} \gamma_{ij}$ and for estimating α, we look at the part in Q which depends on α, which is given by

$$l\,(\alpha) = \sum_{i=1}^{N}\sum_{j=1}^{k} \gamma_{ij} \log f_j\big(x_i | \alpha_j\big)\qquad(7.9)$$

Now, we choose α_j such that $\alpha_j^t = \text{argmax } l(\alpha_j)$, which is obtained by the process of α_j assigning data points to respective clusters, given by argmax γ_{ij}, and estimate α_j by j some estimation method based on the assigned observations to that cluster. It can be seen as a Bayesian concept (although not strictly Bayesian) for learning where equation 7.3 provides the cluster membership probability. In that equation, π_j can be viewed as the prior probability of $Z_i = j$ and the quantity γ_{ij} as the corresponding posterior probability once x is observed. Hence, it is computed using Bayes rule. The idea of choosing the cluster based on maximum probability is the same as choosing the MAP estimate, the mode of the distribution of $Pr(Z_i = j | X, \alpha)$. The MAP estimate is given by

$$Z_i^* = \text{argmax}_j \, \gamma_{ij} = \text{argmax}_j \, \log \, p(x_i | Z_i = j, \alpha) + \log \, p(Z_i = j | \alpha) \quad (7.10)$$

In addition to the general setting of a Hard EM, we include an extra step at the M step of the algorithm as a modification. Instead of obtaining the MLE right away at M step, we propose to do a quality check of the model. For cluster j, $j = 1, \ldots, k$ we denote

$$\alpha_j^{new} = \begin{cases} \alpha_j^{init}, & \text{if cluster } j \text{ is empty} \\ \alpha_j^{MLE}, & \text{otherwise} \end{cases} \quad (7.11)$$

where α_j^{init} is the initial value of the parameter α_j. Hard EM is well known for its greedy convergence; as a result, often, the algorithm converges with one or more clusters being empty. Hence, we would like to force the algorithm to re-iterate if one or more clusters are found to be empty at each M step.

At first, some trial values of the distribution parameters α and mixture proportions π are initialized to start the algorithm. Then the initial value of the log likelihood is evaluated. For different distributions, different techniques can be used to choose suitable initial values. It is known that EM algorithm is very sensitive to the choice of initial values [38]. Hard EM is no different in this regard. However, with proper initialization techniques, Hard EM can provide a robust performance. In the literature, we find different techniques of choosing starting values such as random initialization [39], iteratively constrained EM [40], KMeans clustering [41], Sum scores [42], etc. However, for better performance KMeans initialization and iteratively constrained EM with random initialization are most preferred [43]. There exist some robust versions of EM algorithm (see [44]) which take into account the number of clusters as well. For our study, we have taken, the centroids of KMeans as initial values of μ, and the empirical covariance matrix of each cluster is taken as an initial value of Σ_j. The initial values of π is computed using the ratio of cluster members obtained by KMeans algorithm and total observations.

At the E step, the values of the probabilities γ_{ij} are evaluated using the current parameter values. For a usual EM algorithm (e.g., in a GMM), at the M step, a weighted mean and a weighted covariance matrix are calculated using the γ_{ij} values. But for other distributions, where the model parameters are not mean and (co)variance, this technique can not be used. So, for different distributions, different techniques need to be used. Hard EM provides an easy and convenient solution where at the M step, each data point is assigned to a cluster depending on the probability of that data point belonging to each cluster. That cluster is assigned for which the probability is maximum. Now, if one or more clusters are found empty, then the initial value of the parameter α_j for cluster j is used. And for the non-empty clusters, point estimates of the parameters of each parent distribution are obtained using only the data points available in each cluster. For faster convergence and convenience, maximum likelihood estimates can usually be recommended. The mixture component probabilities π_j are estimated as mentioned above by $\frac{N_j}{N}$. The new set of estimated values of the parameters is then used as an update over the previous one. After this step, the log likelihood is evaluated again using the updated parameter values. The process is then continued until convergence. Hard EM enjoys good convergence properties, which have been explained in detail by [45]. It is to be noted that, although estimates obtained through modified Hard EM can be seen as an approximated MLE and it means that MLE estimates from standard should be better, it is not guaranteed that standard EM would give better accuracy for assignments of data points to correct clusters. As Hard EM finds the MAP estimate i.e., the mode of the distribution of $Pr(Z_i = j|X,\alpha)$ to optimize the classification likelihood, which version of EM gives better accuracy should be investigated for a case-to-case basis.

Algorithm 1: Modified Hard EM Algorithm for Mixture Models

Initialize the model parameters, α and π. Evaluate the initial value of the log likelihood from equation (7.2);

while *loglikelihood difference* $\geq \in$ **do**

    Evaluate γ_{ij} From equation (7.3), using the parameter values and data

    $\pi_j^{new} = \frac{N_j}{N}$, where, $N_j = P_{i=1}^{N} \gamma_{ij}$;

    for *i in1 to N* **do**

        cluster $z_i = \underset{j}{\text{argmax}}\, \gamma_{ij}$;

        Assign data point x_i to cluster Z_i;

    end

    for *j in1 to k* **do**

        if *Cluster j is empty* **then**

            Use initial values of α_j as an update;

        else

            $\alpha_j^{new} = \alpha_j^{MLE}$;

        end

    end

    Re-evaluate log likelihood using the new values of the parameters.

end

For our experiments, we have used 0.0001 as the value of ϵ in Algorithm 1.

In case of a Gaussian mixture model, normal distribution can be used as the base distribution $f_j(.)$ in the model shown in equation 7.1.

For a $p \times 1$ continuous random vector X, the density of p variate multivariate normal distribution is given by

$$f\left(x|\mu, \Sigma\right) = \frac{1}{(2\pi)^{p/2}|\Sigma|^{1/2}} exp\left[-\frac{1}{2}(x - \mu)^T \Sigma^{-1}(x - \mu)\right], \qquad (7.12)$$

where μ is a $p \times 1$ vector, Σ is a $p \times p$ symmetric, positive definite matrix and the support of X is R^p.

The maximum likelihood estimates of μ and Σ are given by

$$\widehat{\mu_{MLE}} = \frac{1}{N} \sum_{i=1}^{N} x_i \qquad (7.13)$$

$$\widehat{\Sigma_{MLE}} = \frac{1}{N} \sum_{i} \left(x_i - \widehat{\mu_{MLE}}\right)\left(x_i - \widehat{\mu_{MLE}}\right)^T \qquad (7.14)$$

The mixture model can be built the usual way and the model parameters can be estimated using the MLEs of a Gaussian distribution at the M step.

7.3 COMPARISON ON BENCHMARK DATA SETS

We have done an extensive experiment to observe the performance of Hard GMM in different conditions. Many authors [46] have argued that it may not be the best idea to test the clustering model only on synthetic data as the model is supposed to perform on real data problems. That is why we have decided to check the performance of our proposed method on both synthetic and real data. In this section, we are going to evaluate the performance of Hard GMM on basic benchmark data sets as proposed by [36]. The data sets are chosen in such a way that the sets are challenging enough for most typical heuristics to fail but easy enough for good clustering algorithms to identify the correct clusters. These data sets have been previously used by many other authors as well. A brief description of the data sets is given below in Table 7.1.

Figure 7.1, Figure 7.2 and Figure 7.3 show plots of data in data set A. In data set A, we have three sets of data containing distinct, separate clusters with the number of clusters 20, 35, and 50, respectively. We have four sets of data containing 15 clusters each in data set S, but with increasing overlap among clusters. Overlap has been increased by increasing the standard deviation of data points in each cluster. It is done in such a way that a good algorithm should still be able to identify the clusters. Figure 7.5, Figure 7.6, Figure 7.7 and Figure 7.8 show plots of data in data set S. Dim data sets contain six sets of data, each with distinct, separate clusters but with

Table 7.1 Descriptions of five basic benchmark data sets

Data set	Variation	Size	Clusters	Dimension	Source
A (3 sets)	Number of Clusters	3000–7500	20,30,50	2	Kärkkäinen, Fränti [47]
S (4 sets)	Overlap	5000	15	2	Fränti, Virmajoki [48]
Dim (6 sets)	Dimensions	1024	16	32–1024	Franti et al., [49]
G2 (100 sets)	Dimensions and overlap	2048	2	2–1024	Fränti et al., [50]
Unbalance (1 set)	Balance	6500	8	2	Rezaei, Fränti [51]

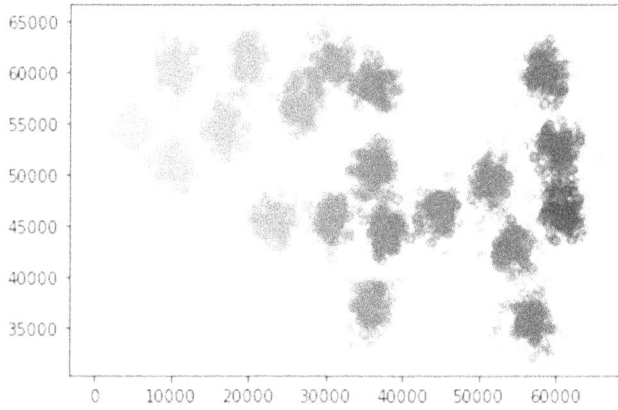

Figure 7.1 A1 data set.

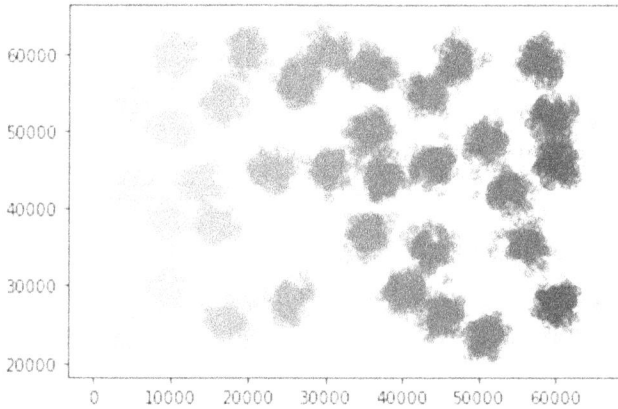

Figure 7.2 A2 data set.

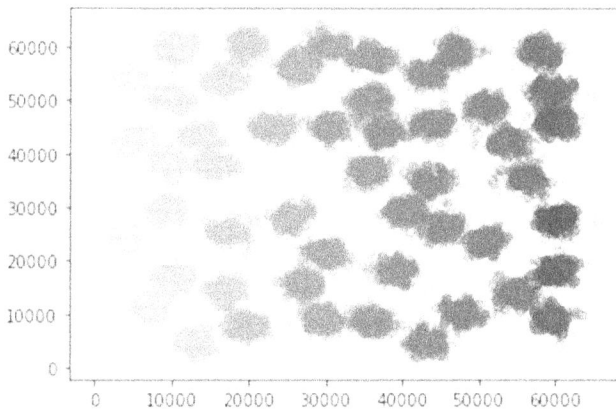

Figure 7.3 A3 data set.

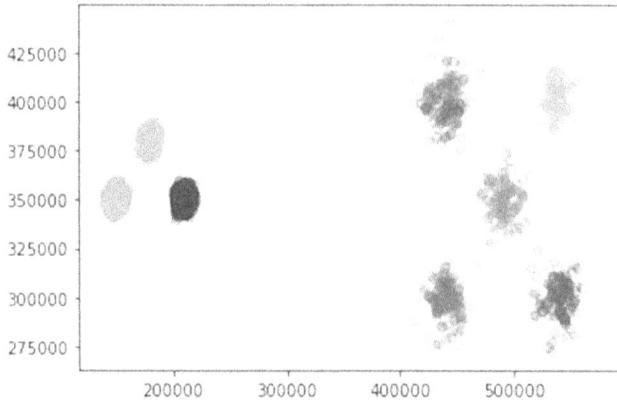

Figure 7.4 Unbalance data set.

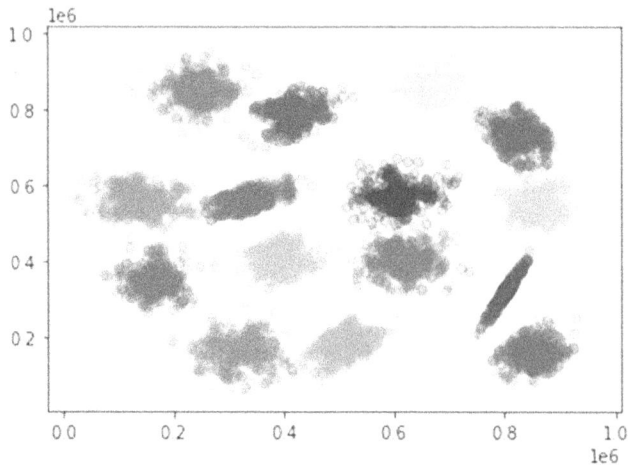

Figure 7.5 SI data set.

increasing dimensions. The data sets have data with dimensions 32, 64, 128, 256, 512 and 1024, respectively. The G2 data sets have 100 sets of data with increasing dimensions and overlap of clusters. With an increase of dimension, the standard deviation of data points of each cluster has also been increased to introduce increasing overlap. The dimension of data sets ranges from 2 to 1024; at the same time, the standard deviation ranges from 10 to 100. For the data set Unbalance, we have eight clusters with an imbalance of data points in each cluster. In other words, a few clusters contain more data points, and a few clusters contain very few data points. Figure 7.4 shows the unbalanced data set.

We have run Hard GMM and Usual GMM algorithms 100 times on each data to measure and compare the mean performance of both models. The

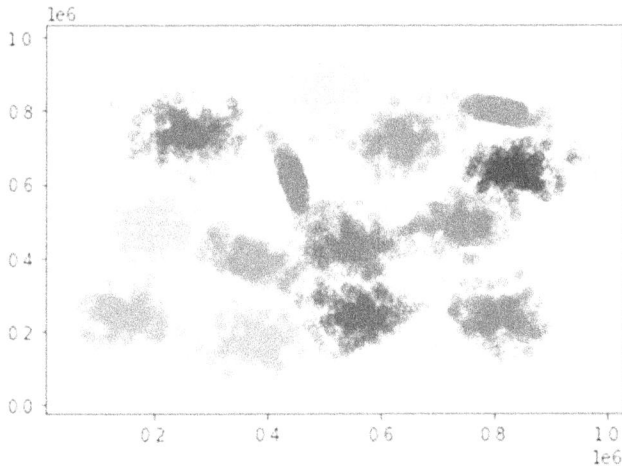

Figure 7.6 S2 data set.

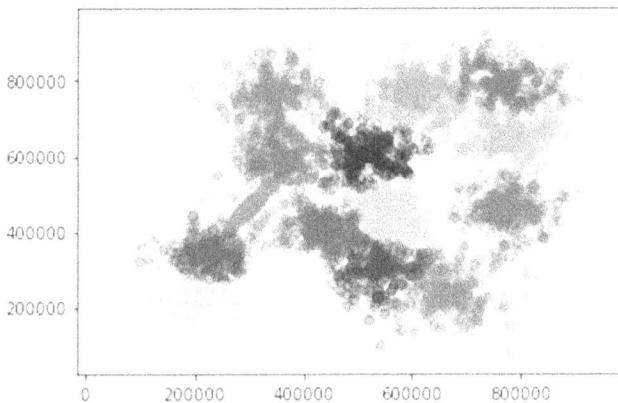

Figure 7.7 S3 data set.

algorithm for usual GMM is available in Scikit-learn [52], a machine learning library in Python. The algorithm for hard GMM is also written in Python. The experiments were done on a machine with 32 gigabytes of RAM and multi-threaded CPU. We have checked three measures to evaluate the performance.

- Accuracy: The total accurate classifications, divided by the number of observations.
- Precision: True positives, divided by sum of true positives and false positives.
- Recall: True positives, divided by the sum of true positives and false negatives.

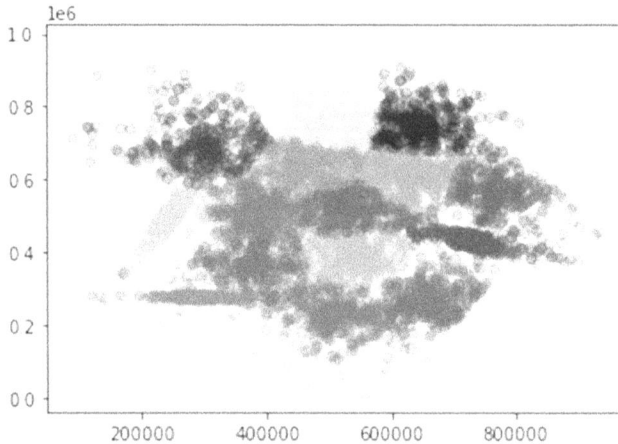

Figure 7.8 S4 data set.

A detailed description of all the measures can be found in [53]. Additionally, we have also provided computational run-time (in seconds) of the models where applicable. The results of the experiments on data sets A and S are given below. For datasets Dim, G2 and unbalance, both Hard GMM and Usual GMM are found to have given 100% success in terms of accuracy, precision and recall. Table 7.2, Table 7.3 and Table 7.4 display the detailed results of data sets A1, A2 and A3, respectively. Whereas the results of data sets S1, S2, S3 and S4 are shown in Table 7.5, Table 7.6, Table 7.7 and Table 7.8.

From the above results in Figure 7.9 and Figure 7.10, we see that Hard GMM works better than Usual GMM on Data sets A and S in terms of accuracy, precision, and recall. We have also observed that Hard GMM is much more consistent for data sets A and S on 100 runs, as the standard deviation is much less for Hard GMM for the data sets A and S. From Figure 7.9 we see that, with an increasing number of clusters, the performance of Hard GMM degrades in terms of accuracy, precision, and recall. However, Usual GMM does not show any significant pattern in performance in terms of accuracy, precision and recall when it comes to the increasing number of clusters in the data. From Figure 7.10 we see that the performance of both Hard GMM and Usual GMM drops significantly with increasing overlap of clusters in terms of accuracy, precision, and recall. But, we have seen no effect of increasing dimension in the performance of both Hard GMM and Usual GMM, as every time over 100 runs, the models have shown 100% success in terms of accuracy, precision, and recall on dim data sets. It is evident that increasing dimension has very little to do in terms of performance if the clusters are distinct. For G2 data sets, we had 100 sets of data with both increasing dimensions and overlap. But, in this case, also, we have observed 100% success of both models to identify the clusters on all 100 runs.

Table 7.2 Performance of Hard GMM and Usual GMM on dataset A1

Measures	AHG	AUG	PHG	PUG	RHG	RUG	RTHG	RTUG
Mean	0.996920	0.875690	0.996878	0.976668	0.996977	0.959366	0.178015	0.077974
Standard Deviation	0.000143	0.116462	0.000151	0.014935	0.000131	0.030430	0.032193	0.007880
Minimum	0.996667	0.626000	0.996610	0.942585	0.996745	0.882558	0.161511	0.062025
First Quartile	0.997000	0.796000	0.996962	0.966382	0.997050	0.938932	0.166006	0.074326
Median	0.997000	0.873333	0.996962	0.972818	0.997050	0.942855	0.168327	0.077453
Third Quartile	0.997000	0.992333	0.996962	0.992125	0.997050	0.992760	0.176055	0.083743
Maximum	0.997000	0.992333	0.996962	0.992125	0.997050	0.992760	0.443339	0.100296

AHG: Accuracy of Hard GMM, AUG: Accuracy of Usual GMM, PHG: Precision of Hard GMM, PUG: Precision of Usual GMM, RHG: Recall of Hard GMM, RUG: Recall of Usual GMM RTHG: Runtime of Hard GMM, RTUG: Run-time of Usual GMM.

Table 7.3 Performance of Hard GMM and Usual GMM on dataset A2

Measures	AHG	AUG	PHG	PUG	RHG	RUG	RTHG	RTUG
Mean	0.977640	0.863086	0.994175	0.977943	0.991563	0.957694	1.502774	0.185822
Standard Deviation	0.045927	0.073965	0.006402	0.008992	0.012301	0.018494	0.152975	0.033539
Minimum	0.783238	0.689143	0.973484	0.951427	0.961736	0.905962	1.356514	0.128678
First Quartile	0.996762	0.807000	0.996731	0.971669	0.996865	0.936981	1.427737	0.167785
Median	0.996952	0.869333	0.996916	0.978609	0.997036	0.963351	1.468551	0.177822
Third Quartile	0.997143	0.915571	0.997116	0.983020	0.997209	0.965244	1.524925	0.193811
Maximum	0.997333	0.993524	0.997305	0.993429	0.997393	0.993822	2.650507	0.394115

AHG: Accuracy of Hard GMM, AUG: Accuracy of Usual GMM, PHG: Precision of Hard GMM, PUG: Precision of Usual GMM, RHG: Recall of Hard GMM, RUG: Recall of Usual GMM RTHG: Runtime of Hard GMM, RTUG: Run-time of Usual GMM.

Table 7.4 Performance of Hard GMM and Usual GMM on dataset A3

Measures	AHG	AUG	PHG	PUG	RHG	RUG	RTHG	RTUG
Mean	0.934595	0.872537	0.990532	0.980345	0.983551	0.960637	1.786347	0.280679
Standard Deviation	0.056556	0.063956	0.005756	0.007250	0.010908	0.015886	0.315283	0.070037
Minimum	0.817733	0.680667	0.980392	0.959247	0.972147	0.909704	1.478938	0.196710
First Quartile	0.889633	0.835833	0.985925	0.976154	0.974674	0.952880	1.556423	0.240818
Median	0.929133	0.877400	0.98208	0.981027	0.97209	0.955084	1.683763	0.259527
Third Quartile	0.997467	0.919100	0.997456	0.985758	0.997513	0.974323	1.891815	0.298878
Maximum	0.997867	0.995333	0.997851	0.995275	0.997906	0.995483	3.232524	0.718938

AHG: Accuracy of Hard GMM, AUG: Accuracy of Usual GMM, PHG: Precision of Hard GMM, PUG: Precision of Usual GMM, RHG: Recall of Hard GMM, RUG: Recall of Usual GMM RTHG: Runtime of Hard GMM, RTUG: Run-time of Usual GMM.

Table 7.5 Performance of Hard GMM and Usual GMM on dataset S1

Measures	AHG	AUG	PHG	PUG	RHG	RUG	RTHG	RTUG
Mean	0.998200	0.974388	0.998186	0.995816	0.998180	0.988442	1.560566	0.165872
Standard Deviation	0.000000	0.063192	0.000000	0.005272	0.000000	0.023293	0.209326	0.041186
Minimum	0.998200	0.715400	0.998186	0.970759	0.998180	0.930539	1.304745	0.093759
First Quartile	0.998200	0.997800	0.998186	0.997787	0.998180	0.997793	1.445687	0.147343
Median	0.998200	0.997800	0.998186	0.997787	0.998180	0.997793	1.483772	0.151386
Third Quartile	0.998200	0.997800	0.998186	0.997787	0.998180	0.997793	1.593771	0.159478
Maximum	0.998200	0.997800	0.998186	0.997787	0.998180	0.997793	2.342927	0.347596

AHG: Accuracy of Hard GMM, AUG: Accuracy of Usual GMM, PHG: Precision of Hard GMM, PUG: Precision of Usual GMM, RHG: Recall of Hard GMM, RUG: Recall of Usual GMM RTHG: Runtime of Hard GMM, RTUG: Run-time of Usual GMM.

Table 7.6 Performance of Hard GMM and Usual GMM on dataset S2

Measures	AHG	AUG	PHG	PUG	RHG	RUG	RTHG	RTUG
Mean	0.992364	0.939530	0.992504	0.980173	0.992340	0.959469	1.678559	0.182416
Standard Deviation	0.000230	0.070728	0.000224	0.007245	0.000231	0.032618	0.378828	0.054915
Minimum	0.991400	0.703600	0.991574	0.950935	0.991350	0.908707	1.369272	0.117463
First Quartile	0.992200	0.908900	0.992336	0.977107	0.992189	0.920079	1.482488	0.157468
Median	0.992400	0.984600	0.992542	0.984818	0.992380	0.984725	1.543244	0.161766
Third Quartile	0.992600	0.984800	0.992733	0.985023	0.992571	0.984927	1.690107	0.180889
Maximum	0.992800	0.984800	0.992953	0.985038	0.992767	0.984927	3.640522	0.440569

AHG: Accuracy of Hard GMM, AUG: Accuracy of Usual GMM, PHG: Precision of Hard GMM, PUG: Precision of Usual GMM, RHG: Recall of Hard GMM, RUG: Recall of Usual GMM RTHG: Runtime of Hard GMM, RTUG: Run-time of Usual GMM.

Table 7.7 Performance of Hard GMM and Usual GMM on dataset S3

Measures	AHG	AUG	PHG	PUG	RHG	RUG	RTHG	RTUG
Mean	0.973476	0.850652	0.973608	0.936191	0.973091	0.905184	1.817396	0.240185
Standard Deviation	0.000627	0.092452	0.000622	0.014524	0.000655	0.038538	0.352305	0.118835
Minimum	0.972200	0.532600	0.972323	0.900290	0.971763	0.774384	1.417350	0.132710
First Quartile	0.973000	0.794100	0.973146	0.924001	0.972592	0.879216	1.532039	0.173736
Median	0.973400	0.846100	0.973554	0.938519	0.973041	0.891337	1.685317	0.206736
Third Quartile	0.974000	0.948000	0.974109	0.949817	0.973637	0.947924	2.026184	0.235066
Maximum	0.974800	0.949800	0.974872	0.951637	0.974501	0.949731	3.048964	0.718240

AHG: Accuracy of Hard GMM, AUG: Accuracy of Usual GMM, PHG: Precision of Hard GMM, PUG: Precision of Usual GMM, RHG: Recall of Hard GMM, RUG: Recall of Usual GMM RTHG: Runtime of Hard GMM, RTUG: Run-time of Usual GMM.

Table 7.8 Performance of Hard GMM and Usual GMM on dataset S4

Measures	AHG	AUG	PHG	PUG	RHG	RUG	RTHG	RTUG
Mean	0.967522	0.846350	0.967441	0.907800	0.966902	0.887625	1.615013	0.230549
Standard Deviation	0.002470	0.066463	0.002502	0.010089	0.002422	0.028414	0.186901	0.046512
Minimum	0.962200	0.709000	0.961969	0.868938	0.961627	0.835819	1.397856	0.174893
First Quartile	0.965400	0.784150	0.965274	0.908700	0.964847	0.863440	1.504002	0.202314
Median	0.967700	0.903000	0.967633	0.909992	0.967221	0.912310	1.564790	0.213907
Third Quartile	0.969600	0.904600	0.969546	0.912384	0.968908	0.913579	1.643767	0.251595
Maximum	0.972000	0.908400	0.972176	0.925609	0.971375	0.916840	2.434432	0.434772

AHG: Accuracy of Hard GMM, AUG: Accuracy of Usual GMM, PHG: Precision of Hard GMM, PUG: Precision of Usual GMM, RHG: Recall of Hard GMM, RUG: Recall of Usual GMM RTHG: Runtime of Hard GMM, RTUG: Run-time of Usual GMM.

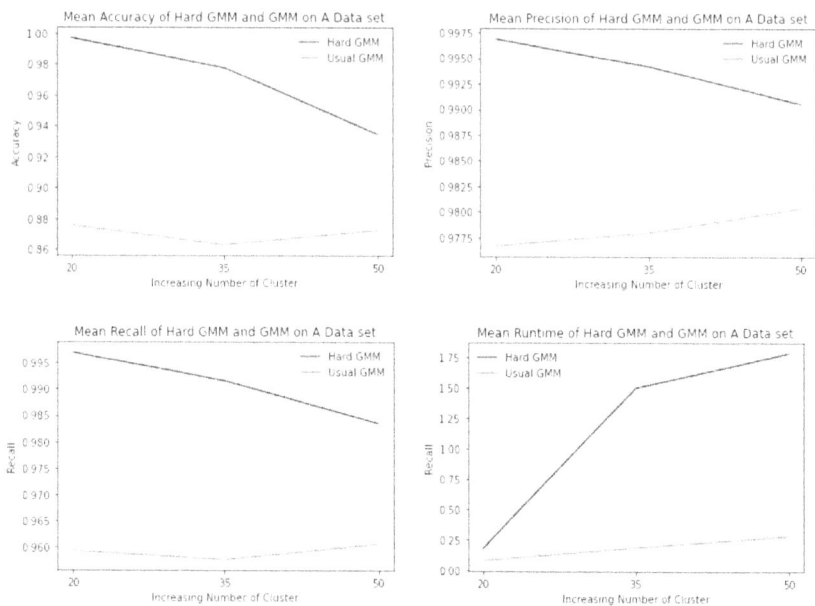

Figure 7.9 Mean accuracy, mean precision, mean recall and mean run-time with increasing number of clusters on A data sets.

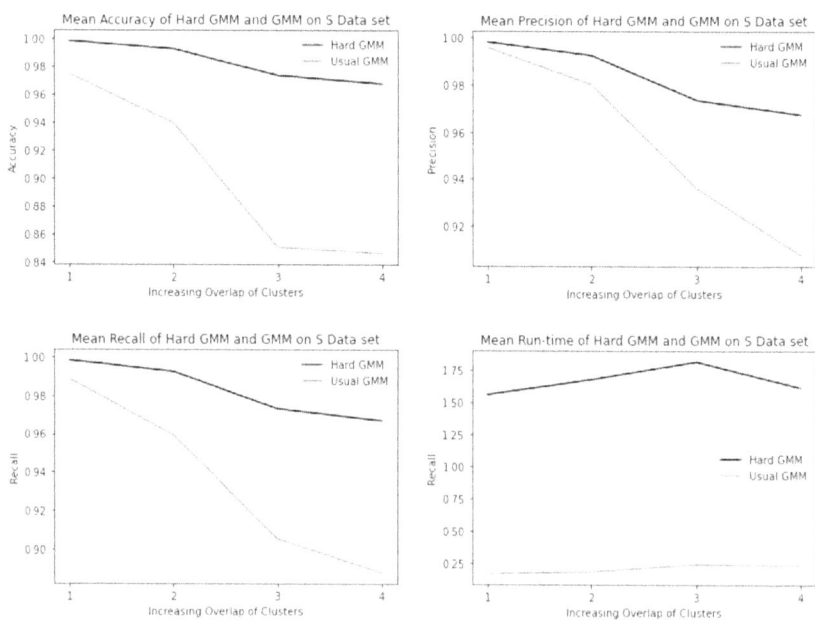

Figure 7.10 Mean accuracy, mean precision, mean recall and mean run-time with increasing overlap of clusters on S data sets.

The result occurs mostly because, in higher dimensional space, the data becomes sparse, and clusters keep on moving far away from each other. Surprisingly, on the unbalanced data set also, we have seen 100% success for both models. It is to be noted that the very popular KMeans algorithm works poorly if the data is imbalanced. Nevertheless, the Gaussian Mixture model, both with hard EM and Usual EM, seems to work very well on imbalanced data. When it comes to computational run time, we see that usual GMM is slightly faster. The clustering strategy of modified Hard EM stops early convergence of the algorithm with non-occupied clusters. From the results, we see that Hard EM offers better results in some situations with a little more run time.

7.4 REAL DATA APPLICATION

We have compared the performance of Hard GMM and Usual GMM on two real data sets. Please note that our study aims to compare the performance of the two methods and not to provide an optimum solution for the problems.

7.4.1 Breast cancer data

The relationship between several risk factors and breast cancer was studied by [54] by assessing hyperresistinemia and metabolic dysregulation in breast cancer. Between 2009 and 2013, women who had been newly diagnosed with breast cancer were recruited from the University Hospital Centre of Coimbra (CHUC). For each patient, the diagnosis was made by positive mammography, and it was histologically confirmed. Before surgery and treatment, all samples were collected, and all the patients with treatment before the consultation were excluded. Healthy female volunteers were selected and enrolled in the study as controls. All patients had no prior cancer treatment, and all participants were free from any infection or other acute diseases or comorbidities at the time of enrollment in the study. Later, [55] also used the data to build supervised learning models and provided an idea for a cheap and effective biomarker for breast cancer. In the dataset, we have nine clinical and biochemical factors, namely: Age (years), BMI (kg/m2),Glucose (mg/dL), Insulin (μU/mL), HOMA, Leptin (ng/mL), Adiponectin (μg/mL), Resistin (ng/mL) and MCP-1(pg/dL). The data can be downloaded from UCI Machine Learning Repository.

7.4.2 Yeast cell cycle data

The fluctuation of expression levels of approximately 6,000 genes over two cell cycles (17-time points) was shown by [56]. Later, [57] studied a subset of 384 genes where they had expression levels peaking at different times corresponding to the five phases of the cell cycle, namely: Early G1, Late

Table 7.9 Performance of Hard GMM and Usual GMM on Real Data

	Breast cancer data	Yeast cell cycle data
Accuracy of Hard GMM	0.543103	0.731771
Accuracy of Usual GMM	0.534482	0.559896
Precision of Hard GMM	0.578726	0.726260
Precision of Usual GMM	0.572716	0.623411
Recall of Hard GMM	0.642159	0.725608
Recall of Usual GMM	0.651250	0.585197
Run-time of Hard GMM	0.016630	0.075423
Run-time of Usual GMM	0.021233	0.076323

The aim of our study is to build a Gaussian Mixture Model using these features and cluster the data points into two categories, namely: healthy controls and patients. In this case, $k = 2$, $p = 9$ and $N = 116$.

G1, S, G2, and M. This data set has been previously used by other authors as well (see [58]). This data is also available at UCI Machine Learning Repository. The aim of the study is to cluster the gene expressions into five categories. In this case $k = 5$, $p = 17$ and $N = 384$.

From the above result in Table 7.9, we see that Hard GMM works better than Usual GMM on both data sets in terms of accuracy and precision. In terms of recall, Usual GMM works better in Breast Cancer data, and Hard GMM works better in cell cycle data. From the results, we understand that the performance of GMM depends mostly on overlap of data in different clusters than a number of clusters or dimensions. In terms of run time, we see that Hard EM was faster than Usual GMM with better results.

7.5 CONCLUSION

In our study, we wanted to address the question if Hard EM really works worse as it is assumed due to its theoretical disadvantages. We also wanted to verify how Hard EM performs with respect to Usual EM in different situations in clustering problems. In our study, we have implemented Hard EM with some modifications, and we have compared the performance of Hard EM and Usual EM on both basic benchmark data sets and real data sets. Furthermore, from our experiments, we have found that it can not be said that Hard EM works worse than Usual EM for clustering analysis. In fact, on many occasions, like an increasing number of clusters and increasing overlap, Hard EM has been found to perform better than Usual EM.

We have already discussed that Hard EM is computationally less intensive, and it is easy to implement, which fulfills the important criteria of choosing a suitable algorithm [59,60]. However, if the limitations of a suitable algorithm are not known, people tend to choose a less accurate algorithm whose limitations are well known beforehand. We have proposed some modifications to Hard EM which stops faster convergence with poor results. Nevertheless, from our experiment we have seen that the difference in run time is not significant. On real data sets the run time of Hard EM was found to be less than that of Usual EM. Moreover, the performance of Hard EM has been enhanced due to the modifications. We have done an extensive experiment on different situations and ran the algorithm on each data 100 times. We have noticed that the performance of Hard EM is very consistent. The robustness of the algorithm can be seen in a varied range of situations. It is often assumed that increasing dimension has an inverse effect on the performance of clustering algorithms [61]. We have found that increasing dimension has almost nothing to do with the performance of Hard GMM if the clusters are distinct and separated. However, proper initialization technique must be observed in order to produce good results.

It is known that KMeans works poorly for imbalanced data [36]. However, for Gaussian Mixture Model, both hard and usual EM are found to work well for imbalanced data points in the cluster. The condition which affects the performance of Hard EM the most is the overlap of clusters. We have seen that the performance drops significantly for both Hard and Usual GMM when the overlap is increased. We have also noticed that for an increasing number of clusters, the performance of Hard GMM drops, whereas Usual GMM shows no significant pattern.

Our study has shown that despite having some disadvantages, Hard EM works at par with Usual GMM. The results on real data sets involving gene expression and biochemical analysis confirm the same. Thus, our study recommends the use of Modified Hard EM for clustering purposes. The proposed model is expected to yield a result at least as good as a standard EM algorithm in the situations we have considered in our study.

REFERENCES

1. Arthur P. Dempster, Nan M. Laird, and Donald B. Rubin. Maximum likelihood from incomplete data via the em algorithm. Journal of the Royal Statistical Society: Series B (Methodological), 39(1):1–22, 1977.
2. Jean-Patrick Baudry and Gilles Celeux. Em for mixtures. Statistics and Computing, 25(4): 713–726, 2015.
3. Geoffrey J. McLachlan, Sharon X. Lee, and Suren I. Rathnayake. Finite mixture models. Annual Review of Statistics and Its Application, 6:355–378, 2019.
4. Jiayi Ma, Xingyu Jiang, Junjun Jiang, and Yuan Gao. Feature-guided Gaussian mixture model for image matching. Pattern Recognition, 92:231–245, 2019.
5. Israel Dejene Gebru, Xavier Alameda-Pineda, Florence Forbes, and Radu Horaud. Em algorithms for weighted-data clustering with application to audio-visual scene analysis. IEEE Transactions on Pattern Analysis and Machine Intelligence, 38(12):2402–2415, 2016.

6. Frederick Jelinek. Continuous speech recognition by statistical methods. Proceedings of the IEEE, 64(4):532–556, 1976.

7. Lawrence Rabiner. Fundamentals of speech recognition. Fundamentals of Speech Recognition, 1993.

8. Hermann Ney, Volker Steinbiss, Reinhold Haeb-Umbach, B-H Tran, and Ute Essen. An overview of the Philips research system for large vocabulary continuous speech recognition. International Journal of Pattern Recognition and Artificial Intelligence, 8(01):33–70, 1994.

9. Biing Hwang Juang and Laurence R. Rabiner. Hidden Markov models for speech recognition. Technometrics, 33(3):251–272, 1991.

10. Nikko Ström, Lee Hetherington, Timothy J. Hazen, Eric Sandness, and James Glass. Acoustic modeling improvements in a segment-based speech recognizer. In Proceedings of IEEE ASRU Workshop, 1999.

11. Volker Steinbiss, Herman Ney, X. Aubert, Stefan Besling, Christian Dugast, Ute Essen, Daryl Geller, Reinhold Haeb-Umbach, Reinhard Kneser, H-G Meier, et al. The Philips research system for continuous-speech recognition. Philips Journal of Research, 49(4):317–352, 1995.

12. Yejin Choi and Claire Cardie. Structured local training and biased potential functions for conditional random fields with application to coreference resolution. In *Human Language Technologies 2007: The Conference of the North American Chapter of the Association for Computational Linguistics; Proceedings of the Main Conference*, pages 65–72, 2007.

13. Mengqiu Wang, Noah A. Smith, and Teruko Mitamura. What is the jeopardy model? a quasisynchronous grammar for qa. In *Proceedings of the 2007 Joint Conference on Empirical Methods in Natural Language Processing and Computational Natural Language Learning (EMNLP-CoNLL)*, pages 22–32, 2007.

14. Sharon Goldwater and Mark Johnson. Representational bias in unsupervised learning of syllable structure. In *Proceedings of the Ninth Conference on Computational Natural Language Learning (CoNLL-2005)*, pages 112–119, 2005.

15. John DeNero and Dan Klein. The complexity of phrase alignment problems. In *Proceedings of ACL-08: HLT, Short Papers*, pages 25–28, 2008.

16. Valentin I. Spitkovsky, Hiyan Alshawi, Dan Jurafsky, and Christopher D. Manning. Viterbi training improves unsupervised dependency parsing. In Proceedings of the Fourteenth Conference on Computational Natural Language Learning, pages 9–17, 2010.

17. Gang Ji and Jeff Bilmes. Backoff model training using partially observed data: Application to dialog act tagging. In *Proceedings of the Human Language Technology Conference of the NAACL, Main Conference*, pages 280–287, 2006.

18. Franz Josef Och and Hermann Ney. Improved statistical alignment models. In *Proceedings of the 38th annual meeting of the association for computational linguistics*, pages 440–447, 2000.

19. Georg B. Ehret, Patrick Reichenbach, Ulrike Schindler, Curt M. Horvath, Stefan Fritz, Markus Nabholz, and Philipp Bucher. DNA binding specificity of different stat proteins: Comparison of in vitro specificity with natural target sites* 210. Journal of Biological Chemistry, 276 (9):6675–6688, 2001.

20. Uwe Ohler, Heinrich Niemann, Guo-chun Liao, and Gerald M. Rubin. Joint modeling of DNA sequence and physical properties to improve eukaryotic promoter recognition. Bioinformatics, 17(suppl 1):S199–S206, 2001.

21. Alexandre Lomsadze, Vardges Ter-Hovhannisyan, Yury O. Chernoff, and Mark Borodovsky. Gene identification in novel eukaryotic genomes by self-training algorithm. Nucleic Acids Research, 33(20):6494–6506, 2005.

22. Dhiraj Joshi, Jia Li, and James Ze Wang. A computationally efficient approach to the estimation of two-and three-dimensional hidden Markov models. IEEE Transactions on Image Processing, 15(7):1871–1886, 2006.

23. Gilles Celeux and G´erard Govaert. A classification em algorithm for clustering and two stochastic versions. Computational Statistics & Data Analysis, 14(3):315–332, 1992.

24. Chris Fraley and Adrian E. Raftery. Model-based clustering, discriminant analysis, and density estimation. Journal of the American Statistical Association, 97(458):611–631, 2002.

25. Radford M. Neal and Geoffrey E. Hinton. A view of the em algorithm that justifies incremental, sparse, and other variants. In Learning in Graphical Models, pages 355–368. Springer, 1998.

26. Stuart Lloyd. Least squares quantization in PCM. IEEE transactions on information theory, 28(2):129–137, 1982.

27. Philip A. Chou, Tom Lookabaugh, and Robert M. Gray. Entropy-constrained vector quantization. IEEE Transactions on acoustics, speech, and signal processing, 37(1):31–42, 1989.

28. Ju¨ri Lember and Alexey Koloydenko. Adjusted Viterbi training. Probability in the Engineering and Informational Sciences, 21(3):451–475, 2007.

29. Armen Allahverdyan and Aram Galstyan. Comparative analysis of Viterbi training and maximum likelihood estimation for hmms. In Advances in Neural Information Processing Systems, pages 1674–1682. Citeseer, 2011.

30. Alexey Koloydenko, Meelis K¨a¨arik, and Ju¨ri Lember. On adjusted Viterbi training. Acta Applicandae Mathematicae, 96(1):309–326, 2007.

31. Brian G. Leroux. Maximum-likelihood estimation for hidden Markov models. Stochastic processes and their applications, 40(1):127–143, 1992.

32. Yariv Ephraim and Neri Merhav. Hidden Markov processes. IEEE Transactions on Information Theory, 48(6):1518–1569, 2002.

33. Z. Hatala and F. Puturuhu. Viterbi algorithm and its application to Indonesian speech recognition. In Journal of Physics: Conference Series, volume 1752, page 012085. IOP Publishing, 2021.

34. Emilie M. Shireman, Douglas Steinley, and Michael J. Brusco. Local optima in mixture modeling. Multivariate Behavioral Research, 51(4):466–481, 2016.

35. Christophe Biernacki, Gilles Celeux, and G´erard Govaert. Choosing starting values for the em algorithm for getting the highest likelihood in multivariate Gaussian mixture models. Computational Statistics & Data Analysis, 41(3–4):561–575, 2003.

36. Pasi Fr¨anti and Sami Sieranoja. K-means properties on six clustering benchmark datasets. Applied Intelligence, 48(12):4743–4759, 2018.

37. Kevin P. Murphy. Machine learning: A probabilistic perspective. MIT Press, 2012.

38. V. Melnykov and I. Melnykov. Initializing the EM algorithm in Gaussian mixture models with an unknown number of components. Computational Statistics & Data Analysis, 56: 1381–1395, 2012.

39. J. Hipp and D. Bauer. Local solutions in the estimation of growth mixture models. Psychological Methods, 11: 36, 2006.

40. G. Lubke and B. Muth´en. Performance of factor mixture models as a function of model size, covariate effects, and class-specific parameters. Structural Equation Modeling: A Multidisciplinary Journal, 14: 26–47, 2007.

41. D. Steinley and M. Brusco. Evaluating mixture modeling for clustering: recommendations and cautions. Psychological Methods, 16: 63, 2011.

42. D. Bartholomew, M. Knott, and I. Moustaki. Latent variable models and factor analysis: A unified approach. John Wiley & Sons, 2011.

43. Shireman, E., Steinley, D., and Brusco, M. Examining the effect of initialization strategies on the performance of Gaussian mixture modeling. Behavior Research Methods, 49: 282–293, 2015, 12, 10.3758/s13428-015-0697-6

44. Yang, M., Lai, C., and Lin, C. A robust EM clustering algorithm for Gaussian mixture models. Pattern Recognition, 45, 3950–3961, 2012.

45. Amke Caliebe and U. Rosler. Convergence of the maximum a posteriori path estimator in hidden Markov models. IEEE Transactions on Information Theory, 48(7): 1750–1758, 2002.

46. Ulrike Von Luxburg, Robert C. Williamson, and Isabelle Guyon. Clustering: Science or art? In Proceedings of ICML workshop on unsupervised and transfer learning, pages 65–79. JMLR Workshop and Conference Proceedings, 2012.

47. Ismo K¨arkk¨ainen and Pasi Fr¨anti. Dynamic local search algorithm for the clustering problem. University of Joensuu Joensuu, Finland, 2002.

48. Pasi Fr¨anti and Olli Virmajoki. Iterative shrinking method for clustering problems. Pattern Recognition, 39(5):761–775, 2006.

49. Pasi Franti, Olli Virmajoki, and Ville Hautamaki. Fast agglomerative clustering using a knearest neighbor graph. IEEE Transactions on Pattern Analysis and Machine Intelligence, 28(11):1875–1881, 2006.

50. Pasi Fr¨anti, Radu Mariescu-Istodor, and Caiming Zhong. Xnn graph. In Joint IAPR International Workshops on Statistical Techniques in Pattern Recognition (SPR) and Structural and Syntactic Pattern Recognition (SSPR), pages 207–217. Springer, 2016.

51. Mohammad Rezaei and Pasi Fr¨anti. Set matching measures for external cluster validity. IEEE Transactions on Knowledge and Data Engineering, 28(8):2173–2186, 2016.

52. F. Pedregosa, G. Varoquaux, A. Gramfort, V. Michel, B. Thirion, O. Grisel, M. Blondel, P. Prettenhofer, R. Weiss, V. Dubourg, J. Vanderplas, A. Passos, D. Cournapeau, M. Brucher, M. Perrot, and E. Duchesnay. Scikit-learn: Machine learning in Python. Journal of Machine Learning Research, 12:2825–2830, 2011.

53. Marina Sokolova and Guy Lapalme. A systematic analysis of performance measures for classification tasks. Information Processing & Management, 45(4):427–437, 2009.

54. Joana Crisostomo, Paulo Matafome, Daniela Santos-Silva, Ana L. Gomes, Manuel Gomes, Miguel Patr´ıcio, Liliana Letra, Ana B. Sarmento-Ribeiro, Lelita Santos, and Raquel Sei¸ca. Hyperresistinemia and metabolic dysregulation: A risky crosstalk in obese breast cancer. Endocrine, 53(2):433–442, 2016.

55. Miguel Patrĭcio, Josée Pereira, Joana Crisóstomo, Paulo Matafome, Manuel Gomes, Raquel Seiça, and Francisco Caramelo. Using resistin, glucose, age and BMI to predict the presence of breast cancer. BMC Cancer, 18(1):1–8, 2018.

56. Raymond J. Cho, Michael J. Campbell, Elizabeth A. Winzeler, Lars Steinmetz, Andrew Conway, Lisa Wodicka, Tyra G. Wolfsberg, Andrei E. Gabrielian, David Landsman, David J. Lockhart, et al. A genome-wide transcriptional analysis of the mitotic cell cycle. Molecular Cell, 2(1): 65–73, 1998.

57. Ka Yee Yeung, Chris Fraley, Alejandro Murua, Adrian E. Raftery, and Walter L. Ruzzo. Model-based clustering and data transformations for gene expression data. Bioinformatics, 17(10):977–987, 2001.

58. Zhe Liu, Yu-qing Song, Cong-hua Xie, and Zheng Tang. A new clustering method of gene expression data based on multivariate Gaussian mixture models. Signal, Image and Video Processing, 10(2):359–368, 2016.

59. Tomi Kinnunen, Ilja Sidoroff, Marko Tuononen, and Pasi Fränti. Comparison of clustering methods: A case study of text-independent speaker modeling. Pattern Recognition Letters, 32(13):1604–1617, 2011.

60. Xiaojuan Huang, Li Zhang, Bangjun Wang, Fanzhang Li, and Zhao Zhang. Feature clustering based support vector machine recursive feature elimination for gene selection. Applied Intelligence, 48(3):594–607, 2018.

61. Carlotta Domeniconi, Dimitrios Gunopulos, Sheng Ma, Bojun Yan, Muna Al-Razgan, and Dimitris Papadopoulos. Locally adaptive metrics for clustering high dimensional data. Data Mining and Knowledge Discovery, 14(1):63–97, 2007.

Chapter 8

Impatient customers on an M/M/1 queueing system subjected to differentiated vacations

M. I. G. Suranga Sampath, K. Kalidass, Jicheng Liu, Rakesh Kumar, and Bhupender Kumar Som

8.1 INTRODUCTION

With the gradual increase of interest in research on queueing systems with vacations in the last decades, many authors have been paying attention to the queueing models subjected to server vacations. Applications of queueing systems with server vacations can be observed in many real world applications such as communication networks, flexible manufacturing systems, air traffic control and computer systems. A server in a queueing system may not be available for a random period of time due to factors such as the supply of service to arrivals as in priority queueing discipline, failure of the server by external effects or some other reasons, inability of humans to work continuously without rest, preventive maintenance period in a production system and secondary tasks assigned to the server.

Queueing systems follow a number of vacation policies such as single vacation, multiple vacation and working vacation. In the single vacation policy, a server is allowed to take a vacation of a random period of time when no customers are waiting for the service. After completing a vacation, the server is available in the queueing system to provide the service for the next customers. Otherwise, the server has to wait for incoming customers. A little difference can be observed between the single and multiple vacation policies. In a single vacation policy whether there are waiting customers or not the server returns to the busy period. Whereas in the multiple vacation policy, the server immediately takes another vacation if there are no customers and it returns to the system after completing a vacation. The server continuously serves with slower service rate without completely stopping its services during a vacation in the working vacation policy. Levy and Yechiali [1] are the first to discuss the aspiration of queueing systems with server vacations. The survey by Tian and Zhang [2] and the books by Doshi [3] and Takagi [4] may guide interested readers to know more about the vacation models. Servi and Finn [5] studied the queueing system with working vacation. Their M/M/1/WV model was extended by Wu and Takagi [6] for developing an M/G/1/WV model and it was also extended by Baba [7] to a G/M/1/WV queueing system.

DOI: 10.1201/9781003356653-8

Most authors focused on deriving the steady-state or equilibrium performances for vacation queueing models, since calculating the transient state probabilities is considered to be more complicated than that of simpler cases. Queueing systems that never reach equilibrium are not described by the time independent measures. And the results which are obtained for the steady-state are not applicable in situations which have finite operational time horizon. The behavior of the queueing system when the concerned parameters of the system are perturbed are explained by transient analysis. This helps to study the finite time properties and to obtain the optimal solutions for controlling the queueing systems.

Arumuganathan and Jeyakumar [8] discussed an N-policy single server queue with multiple server vacation and close down times. Kalidass et al. [9] derived the transient state result of a single server queueing system with multiple vacation policy and the possibility of the occurrence of catastrophes. Also, Kalidass and Ramanath [10] investigated an M/M/1 queue with server vacations and a waiting server and discussed the time-dependent result. Indra and Renu [11] derived the transient solution for two dimensional single server queueing model subjected to Bernoulli schedule together with working vacation policy. Sudhesh and Raj [12] derived the time-dependent result for a number of customers in a single server queueing system where the server was allowed to take multiple vacations. The time dependent result of an M/M/1/N queue which has multiple vacation policy and subjected to working breakdowns has been investigated by Yang and Wu [13]. Kalidass and Ramanath [14] discussed the transient behavior of a single server queue with multiple vacations. Vijayashree and Janani [15] studied an N-policy single server queue with multiple exponential vacations and explicit transient solution of system size probabilities was derived by using Laplace transform and probability generating function techniques. Gahlawat et al. [16] obtained the transient result for a two state time-dependent bulk queue model.

Recently, Vijayashree and Janani [17] extended a previous research of Ibe and Isijola [18] to obtain the corresponding steady-state system size probabilities of a queueing model in which a new type of vacation called differentiated vacation was introduced. The authors derived a transient solution for system size probabilities in terms of a modified Bessel function of the first kind using generating function and Laplace transform techniques. Some performance measures such as the mean and variance of the number of customers in the system as well as the probability of the server being in a busy or vacation state at an arbitrary time were derived. The queues subjected to differentiated vacation have two types of vacations; the first takes longer duration vacation and the second type is allowed to take a shorter duration. The longer duration vacation is taken by the server after finishing all the services while it is allowed to take a shorter duration vacation if the system is empty when it returns from the previous vacation. This type of vacations can be observed in the behavior of many real life service workers such as an assistant in a library, a clerk in a bank, an attendant in a gas station and a cashier in a supermarket. They serve the arriving customers and take a scheduled vacation of longer duration. If there are no more customers in the system, they may enter another vacation of shorter duration.

The concept of customer impatience is one of the most important characteristics of a customer in queueing systems. While waiting for service, the customers usually feel anxious and impatient. The waiting customers abandon the queueing system if the server in the system is in idle and the customer's impatient time expires before the server restarts the service. The abandonment of the queueing system by customers affects the stability and performance of queueing models in daily life such as call centers, communication networks and production-inventory systems. Therefore, in the past few decades, many authors have paid attention in studying queueing systems with customers' impatience and server vacations.

Altman and Yechiali [19] investigated the behavior of impatient customers of M/M/1, M/G/1 and M/M/c queues under server vacations where a customer becomes impatient when the servers enter a vacation. The authors analyzed both single and multiple vacation policies and transient solutions were obtained for both cases. The infinite capacity queue subjected to server vacations together with customers' impatience was studied by Altman and Yechiali [20]. The explicit expression of system size probabilities was obtained by using generating function method and the key performance measures were derived by the authors. Perel and Yechiali [21] studied a multiple server queueing model in a two-phase (fast and slow) Markovian random environment by considering the effect of customers' impatience. Also, a single server queueing system with customers' impatience and working vacation was investigated by Yue et al. [22]. They obtained the time-dependent solution for a number of jobs in the queueing system by making use of the probability generating function techniques when the server takes a working vacation. Ammar [23] obtained the transient solution for a single server queue with customers' impatience and multiple vacation policy. Explicit expression as well as the mean and variance for the number of units in the queueing system were derived in terms of the modified Bessel functions of the first kind. He used the probability generating functions with the mathematical treatments of continued fractions together with some properties of the confluent hyper-geometric function. Also, Ammar [24] used similar mathematical calculations to derive a new explicit solution for an M/M/1 vacation queue with impatient customers and a waiting server, where the server is allowed to take a vacation whenever the system is empty after waiting for a random period of time.

Impatient customers on queueing systems subjected to differentiated vacations can be used to explain the behavior of many physical systems. In our model, a hospital emergency room with impatient customers is considered as an example. After some operations, the emergency room must be set up for next patients, but this takes some random time duration since the equipment have to be ready and the room has to be cleaned up and sterilized. This scenario is referred as type 1 vacation. During type 2 vacation, the emergency room personnel take some actual rest while the room is being set up and cleaned. When the system is in any type of vacation, a critical patient may be transferred to another location according to the patients' condition. This situation can be considered as customer abandonments of the system. Another example is the IEEE 802.16e power saving mechanism applied in recent mobile technologies such as WiFi, 3G and WiMAX. When the mobile device has no requests to be served, it goes to power save/sleep mode in order to reduce power consumption. This scenario corresponds to vacation of the server. During a vacation, a message may be recalled by the sender. This can be compared to a similar situation involving an impatient customer in the system.

After reviewing the literature related to the above cases, it was identified that no author tried to derive transient solutions for impatient behavior of customers in a single server queueing system subjected to differentiated server vacations. Thus, the most important task of this research is to develop the transient solutions of an M/M/1 queue by taking customers' impatience and differentiated vacations into account. This research uses the techniques of generating functions and Laplace transform along with some known results of continued fractions and some properties of the confluent hyper-geometric function. This can be considered as a further development of the earlier results which were derived by Vijayashree and Janani [17], by adding the concept of impatient customers.

Sections of this chapter are organized as follows. The model is presented in section 8.2. The explicit expression for the time-dependent system size probabilities is derived in section 8.3. Section 8.4 presents the explicit expressions for time-dependent expected system size and variance. Numerical illustrations are given in section 8.6 and the paper is concluded in section 8.7.

8.2 MODEL DESCRIPTION

As the model, we consider a single server Markovian queueing system subjected to differentiated vacations and impatient customers. The following assumptions are established for the model:

I. Customers join the queueing system following a Poisson process with rate λ and the service rate is exponentially distributed with rate μ.

II. The server can go to take a vacation (type 1) after fulfilling all the service requirements of customers. It must return if at least one arrival is waiting to receive service, otherwise the server takes another vacation (type 2). The two vacation policies follow exponential distributions with parameter γ_1 and γ_2, respectively.

III. Customers waiting in the queue become impatient when the system is taking a vacation. That is, each customer, upon arrival, activates an individual timer, exponentially distributed with parameters ξ_1 and ξ_2 for type-1 and type-2 vacations, respectively such that, if the customer's service has not been started before the customer's impatience timer expires, he/she abandons the system never to return.

IV. It is assumed that inter-arrival, service and vacation times are mutually independent and the scheduling discipline is First-In, First-Out (FIFO).

Let $\{I(t), t \geq 0\}$ denote the total number of customers in the system at time t and let $G(t)$ represent the state of the system at time t, which is defined as follows:

$$G(t) = \begin{cases} 0, & \text{if the server being in functional state at time } t \\ 1, & \text{if the server being in type} - 1 \text{ vacation at time } t \\ 2, & \text{if the server being in type} - 2 \text{ vacation at time } t \end{cases}$$

Then $G(t), I(t), t \geq 0\}$ is a two-dimensional continuous time Markov process on the state space $S = \{(g, n); g = 0, 1, 2; n = 0, 1, 2, 3, 4 \ldots\}$. Let $P_{g,n}(t)$ be the time-dependent probabilities for the system to be in the state g with n customers at time t. Let

$P_{0,n}(t) = Prob\{G(t) = 0, I(t) = n\}, n = 1, 2, \ldots$

$P_{1,n}(t) = Prob\{G(t) = 1, I(t) = n\}, \quad n = 0, 1, 2, \ldots$

$P_{2,n}(t) = Prob\{G(t) = 2, I(t) = n\}, \quad n = 0, 1, 2, \ldots$

Then, the set of forward Kolmogorov differential-difference equations governing the process are given by

$$P'_{0,1}(t) = -(\lambda + \mu)P_{0,1}(t) + \mu P_{0,2}(t) + \gamma_1 P_{1,1}(t) + \gamma_2 P_{2,1}(t), \quad n = 1 \quad (8.1)$$

$$P'_{0,n}(t) = \lambda P_{0,n-1}(t) - (\lambda + \mu)P_{0,n}(t) + \mu P_{0,n+1}(t) + \gamma_1 P_{1,n}(t) + \gamma_2 P_{2,n}(t), 2 \leq n \quad (8.2)$$

$$P'_{1,0}(t) = -\left(\lambda + \gamma_1\right)P_{1,0}(t) + \mu P_{0,1}(t) + \xi_1 P_{1,1}(t), \quad n = 0 \quad (8.3)$$

$$P'_{1,n}(t) = \lambda P_{1,n-1}(t) - \left(\lambda + n\xi_1 + \gamma_1\right)P_{1,n}(t) + (n + 1)\xi_1 P_{1,n+1}(t), \quad 1 \le n$$

$$(8.4)$$

$$P'_{2,0}(t) = -\lambda P_{2,0}(t) + \xi_2 P_{2,1}(t) + \gamma_1 P_{1,0}(t), \quad n = 0 \qquad (8.5)$$

$$P'_{2,n}(t) = \lambda P_{2,n-1}(t) - \left(\lambda + n\xi_2 + \gamma_2\right)P_{2,n}(t) + (n + 1)\xi_2 P_{2,n+1}(t), \quad 1 \le n$$

$$(8.6)$$

It is assumed that zero customers are in the queueing system at time $t = 0$ and the server is in type-1 vacation at that time, i.e., $P_{0,1}(t) = 1$, $P_{2,0}(t) = 0$ and $P_{g,n}(t) = 0$ for $n \ge 1$ and $g = 0, 1, 2$.

8.3 TRANSIENT PROBABILITIES

8.3.1 Evaluation of $P_{1,n}(t)$ and $P_{2,n}(t)$

Let $\hat{P}_{j,n}(s)$ be the Laplace transform of $P_{j,n}(t)$; $n = 1, 2, 3 \ldots$. On taking the Laplace transform of equations (8.4) and (8.6), we have

$$s\hat{P}_{1,n}(s) - P_{1,n}(0) = \lambda\hat{P}_{1,n-1}(s) - \left(\lambda + n\xi_1 + \gamma_1\right)\hat{P}_{1,n}(s) + (n + 1)\xi_1\hat{P}_{1,n+1}(s),$$

$$(8.7)$$

$$s\hat{P}_{2,n}(s) - P_{2,n}(0) = \lambda\hat{P}_{2,n-1}(s) - \left(\lambda + n\xi_2 + \gamma_2\right)\hat{P}_{2,n}(s) + (n + 1)\xi_2\hat{P}_{2,n+1}(s).$$

$$(8.8)$$

Applying boundary conditions to equation (8.7), we have the expression

$$\frac{\hat{P}_{1,n}(s)}{\hat{P}_{1,n-1}(s)} = \frac{\lambda}{\left(s + \lambda + n\xi_1 + \gamma_1\right) - (n + 1)\dfrac{\hat{P}_{1,n+1}(s)}{\hat{P}_{1,n}(s)}}$$

After rearranging the above equation, it will become the following form

$$\frac{\hat{P}_{1,n}(s)}{\hat{P}_{1,n-1}(s)} = \frac{\lambda}{\xi_1\left(\frac{s+\gamma_1}{\xi_1} + n\right)} \cdot \cfrac{\left(\frac{s+\gamma_1}{\xi_1} + n\right)}{\left[\frac{s+\gamma_1}{\xi_1} + n - \left(-\frac{\lambda}{\xi_1}\right) + \cfrac{(n+1)\left(-\frac{\lambda}{\xi_1}\right)}{\left[\frac{(s+\gamma_1)}{\xi_1} + (n+1) - \left(-\frac{\lambda}{\xi_1}\right) + \cfrac{(n+2)\left(-\frac{\lambda}{\xi_1}\right)}{\left[\frac{(s+\gamma_1)}{\xi_1} + (n+2) + \ldots\right]}\right]}\right]}$$

Again, this equation can be rearranged using the identity (B.2) of confluent hyper-geometric function. Then it becomes the following form

$$\frac{\hat{P}_{1,n}(s)}{\hat{P}_{1,n-1}(s)} = \frac{\lambda}{\xi_1\left(\frac{s+\gamma_1}{\xi_1}+n\right)} \frac{{}_1F_1\left(n+1; \frac{s+\gamma_1}{\xi_1}+n+1; -\frac{\lambda}{\xi_1}\right)}{{}_1F_1\left(n; \frac{s+\gamma_1}{\xi_1}+n; -\frac{\lambda}{\xi_1}\right)}$$

For $n = 1, \ 2, \ 3, \ \ldots$ invoking the above equation, we will have

$$\hat{P}_{1,n}(s) = \left(\frac{\lambda}{\xi_1}\right)^n \frac{1}{\prod_{i=1}^{n}\left(\frac{s+\gamma_1}{\xi_1}+i\right)} \frac{{}_1F_1\left(n+1; \frac{s+\gamma_1}{\xi_1}+n+1; -\frac{\lambda}{\xi_1}\right)}{{}_1F_1\left(n; \frac{s+\gamma_1}{\xi_1}+n; -\frac{\lambda}{\xi_1}\right)}\hat{P}_{1,0}(s) \quad (8.9)$$

let

$$\hat{\phi}_n(s) = \left(\frac{\lambda}{\xi_1}\right)^n \frac{1}{\prod_{i=1}^{n}\left(\frac{s+\gamma_1}{\xi_1}+i\right)} \frac{{}_1F_1\left(n+1; \frac{s+\gamma_1}{\xi_1}+n+1; -\frac{\lambda}{\xi_1}\right)}{{}_1F_1\left(n; \frac{s+\gamma_1}{\xi_1}+n; -\frac{\lambda}{\xi_1}\right)}$$

Then

$$P_{1,n}(t) = \phi_n(t) * P_{1,0}(t) \quad (8.10)$$

where $\phi_n(t)$ is the inverse Laplace transform of $\hat{\phi}_n(s)$ and '$*$' denotes convolution.

Applying boundary conditions in equation (8.8) and by using the same procedure which was used to evaluate $P_{1,n}(t)$, we can find $P_{2,n}(t)$ as follows

$$P_{2,n}(t) = \psi_n(t) * P_{2,0}(t) \quad (8.11)$$

where $\psi_n(t)$ is the inverse Laplace transform of $\hat{\psi}_n(s)$ and '$*$' denotes convolution. Also,

$$\hat{\psi}_n(s) = \left(\frac{\lambda}{\xi_2}\right)^n \frac{1}{\prod_{i=1}^{n}\left(\frac{s+\gamma_2}{\xi_2}+i\right)} \frac{{}_1F_1\left(n+1; \frac{s+\gamma_2}{\xi_2}+n+1; -\frac{\lambda}{\xi_2}\right)}{{}_1F_1\left(n; \frac{s+\gamma_2}{\xi_2}+n; -\frac{\lambda}{\xi_2}\right)}$$

8.3.2 Evaluation of $P_{0,n}(t)$

Define the generating function as follows, for $|z| \le 1$

$$P(z, t) = \sum_{n=1}^{\infty} P_{0,n}(t) z^n$$

with initial condition $P(z, 0) = 0$.

Multiplying equations (8.1) and (8.2) by appropriate powers of z and summing over $n \geq 1$, we can obtain

$$\frac{\partial P(z, t)}{\partial t} = \sum_{n=1}^{\infty} P'_{0,n}(t) z^n$$

$$\frac{\partial P(z, t)}{\partial t} + [\lambda(1 - z) + \mu(1 - z^{-1})] P(z, t)$$

$$= \gamma_1 \sum_{n=1}^{\infty} P_{1,n}(t) z^n + \gamma_2 \sum_{n=1}^{\infty} P_{2,n}(t) z^n - \mu P_{0,1}(t)$$

Since the above equation is a first-order partial differential equation for $P(z, t)$, after solving it

using the integrating factor $\exp\{[\lambda(1 - z) + \mu(1 - z^{-1})]t\}$, we will have

$$P(z, t) = \gamma_1 \int_0^t \left(\sum_{m=1}^{\infty} P_{1,m}(u) z^m \right) e^{-[\lambda(1-z)+\mu(1-z^{-1})](t-u)} du$$

$$+ \gamma_2 \int_0^t \left(\sum_{m=1}^{\infty} P_{2,m}(u) z^m \right) e^{-[\lambda(1-z)+\mu(1-z^{-1})](t-u)} du \qquad (8.12)$$

$$- \mu \int_0^t P_{0,1}(u) e^{-[\lambda(1-z)+\mu(1-z^{-1})](t-u)} du$$

It is well known that if $\alpha = 2\sqrt{\lambda\mu}$ and $\beta = \sqrt{\frac{\lambda}{\mu}}$, then

$$exp\left[\left(\lambda z + \frac{\mu}{z} \right) t \right] = \sum_{n=-\infty}^{\infty} (\beta z)^n I_n(\alpha t)$$

where $I_n(.)$ is the modified Bessel function of the first kind.

Comparing the coefficients of z^n of equation (8.12), $n = 1, 2, 3, \ldots.$, on both sides, we will have

$$P_{0,n}(t) = \gamma_1 \int_0^t \sum_{m=1}^{\infty} P_{1,m}(u) \beta^{n-m} I_{n-m}(\alpha(t - u)) e^{-(\lambda+\mu)(t-u)} du$$

$$+ \gamma_2 \int_0^t \sum_{m=1}^{\infty} P_{2,m}(u) \beta^{n-m} I_{n-m}(\alpha(t - u)) e^{-(\lambda+\mu)(t-u)} du \qquad (8.13)$$

$$- \mu \int_0^t P_{0,1}(u) \beta^n I_n(\alpha(t - u)) e^{-(\lambda+\mu)(t-u)} du$$

Using the fact that $I_{-n}(t) = I_n(t)$ and comparing the coefficients of z^{-n} on both sides of equation (8.12), we have

$$0 = \gamma_1 \int_0^t \sum_{m=1}^{\infty} P_{1,m}(u) \beta^{n-m} I_{n+m}(\alpha(t-u)) e^{-(\lambda+\mu)(t-u)} du$$

$$+ \gamma_2 \int_0^t \sum_{m=1}^{\infty} P_{2,m}(u) \beta^{n-m} I_{n+m}(\alpha(t-u)) e^{-(\lambda+\mu)(t-u)} du \qquad (8.14)$$

$$- \mu \int_0^t P_{0,1}(u) \beta^n I_n(\alpha(t-u)) e^{-(\lambda+\mu)(t-u)} du$$

Subtracting equation (8.13) from equation (8.14) for $n = 1, 2, 3, \ldots$, we have

$$P_{0,n}(t) = \gamma_1 \int_0^t \sum_{m=1}^{\infty} P_{1,m}(u) \beta^{n-m} [I_{n-m}(\alpha(t-u)) - I_{n+m}(\alpha(t-u))] e^{-(\lambda+\mu)(t-u)} du$$

$$+ \gamma_2 \int_0^t \sum_{m=1}^{\infty} P_{2,m}(u) \beta^{n-m} [I_{n-m}(\alpha(t-u)) - I_{n+m}(\alpha(t-u))] e^{-(\lambda+\mu)(t-u)} du$$

$$(8.15)$$

Taking the Laplace transform of the above equation and substituting equations (8.10) and (8.11) in it, we can derive

$$\hat{P}_{0,n}(s) = \gamma_1 \frac{\hat{P}_{1,0}(s)}{\sqrt{p^2 - \alpha^2}} \sum_{m=1}^{\infty} \beta^{n-m} \hat{\phi}_m(s) \left\{ \frac{\left(p - \sqrt{p^2 - \alpha^2}\right)^{n-m}}{\alpha^{n-m}} - \frac{\left(p - \sqrt{p^2 - \alpha^2}\right)^{n+m}}{\alpha^{n+m}} \right\}$$

$$+ \gamma_2 \frac{\hat{P}_{2,0}(s)}{\sqrt{p^2 - \alpha^2}} \sum_{m=1}^{\infty} \beta^{n-m} \hat{\psi}_m(s) \left\{ \frac{\left(p - \sqrt{p^2 - \alpha^2}\right)^{n-m}}{\alpha^{n-m}} - \frac{\left(p - \sqrt{p^2 - \alpha^2}\right)^{n+m}}{\alpha^{n+m}} \right\}$$

where $p = s + \lambda + \mu$. The inversion of above equation will yield an explicit expression for $P_{0,n}(t)$ as follows:

$$P_{0,n}(t) = \gamma_1 P_{1,0}(t) * \sum_{m=1}^{\infty} \beta^{n-m} \phi_m(t) * [I_{n-m}(\alpha(t-u))$$

$$- I_{n+m}(\alpha(t-u))] e^{-(\lambda+\mu)t}$$

$$+ \gamma_2 P_{2,0}(t) * \sum_{m=1}^{\infty} \beta^{n-m} \psi_m(t) * [I_{n-m}(\alpha(t-u))$$

$$- I_{n+m}(\alpha(t-u))] e^{-(\lambda+\mu)t}$$

$$(8.16)$$

where '*' denotes convolution. Now, it is clear that $P_{0,n}(t)$ are expressed in terms of $P_{1,0}(t)$ and $P_{2,0}(t)$. They are given by equations (8.18) and (8.20) respectively, for $n = 1, 2, 3, \ldots$.

8.3.3 Evaluation of $P_{2,0}(t)$

Taking Laplace transform of equation (8.5) and applying boundary conditions, we have

$$\hat{P}_{2,0}(s) = \frac{\xi_2}{(s + \lambda)}\hat{P}_{2,1}(s) + \frac{\gamma_1}{(s + \lambda)}\hat{P}_{1,0}(s)$$

Substituting equation (8.11) when $n = 1$ in the above equation, we can obtain

$$\hat{P}_{2,0}(s) = \gamma_1 \sum_{r=0}^{\infty} \frac{\xi_2^r \hat{\psi}_1^r(s)}{(s + \lambda)^{r+1}}\hat{P}_{1,0}(s)$$

Taking inverse Laplace transform of the above equation, we have

$$P_{2,0}(t) = \gamma_1 \sum_{r=0}^{\infty} \xi_2^r e^{-\lambda t}\frac{t^r}{r!} * \psi_1^r(t) * P_{1,0}(t) \tag{8.17}$$

It is clear that $P_{2,0}(t)$ is expressed in terms of $P_{1,0}(t)$. It is given by equation (8.19).

8.3.4 Evaluation of $P_{1,0}(t)$

Taking Laplace transform of equation (8.3) with boundary conditions, we have

$$\hat{P}_{2,0}(s) = \frac{1}{\left(s + \lambda + \gamma_1\right) - \xi_1\frac{\hat{P}_{1,1}(s)}{\hat{P}_{1,0}(s)} - \mu\frac{\hat{P}_{0,1}(s)}{\hat{P}_{1,0}(s)}} \tag{8.18}$$

Applying $n = 1$ in equation (8.15) and after some algebra, we can obtain

$$P_{0,1}(t) = 2\gamma_1 \int_0^t \sum_{m=1}^{\infty} m\beta^{1-m}P_{1,m}(u)\frac{I_m(\alpha(t-u))}{\alpha(t-u)}e^{-(\lambda+\mu)(t-u)}du$$
$$+ 2\gamma_2 \int_0^t \sum_{m=1}^{\infty} m\beta^{1-m}P_{2,m}(u)\frac{I_m(\alpha(t-u))}{\alpha(t-u)}e^{-(\lambda+\mu)(t-u)}du$$

Taking Laplace transform of the above equation and substituting equations (8.10) and (8.11) in it, we will have

$$\hat{P}_{0,1}(s) = 2\gamma_1 \sum_{m=1}^{\infty} \beta^{1-m} \hat{\phi}_m(s) \hat{P}_{1,0}(s) \frac{\alpha^{m-1}}{\left(p + \sqrt{p^2 - \alpha^2}\right)^m}$$

$$+ 2\gamma_2 \sum_{m=1}^{\infty} \beta^{1-m} \hat{\psi}_m(s) \hat{P}_{2,0}(s) \frac{\alpha^{m-1}}{\left(p + \sqrt{p^2 - \alpha^2}\right)^m}$$

where $p = s + \lambda + \mu$.

Substituting the above equation and equation (8.10) when $n = 1$ in equation (8.18) and after some algebra, we can obtain

$$\hat{P}_{0,1}(s) = \sum_{k=0}^{\infty} \sum_{i=0}^{k} \sum_{j=0}^{k-i} (-1)^{i+j} \gamma_1^k \gamma_2^{k-i-j} \binom{k}{i}\binom{k-i}{j}\left(\frac{\xi_1}{\gamma_1}\right)^i \frac{\hat{\phi}_{1(s)}^i \hat{\phi}_{m(s)}^{k-i} \hat{\psi}_{m(s)}^{k-i-j}}{\left(s + \lambda + \gamma_1\right)^{r+1}}$$

$$\times \left[\sum_{m=1}^{\infty}\left(\frac{p - \sqrt{p^2 - \alpha^2}}{\alpha\beta}\right)^m\right]^{k-i} \left[\sum_{r=0}^{\infty}\frac{\xi_2^r \hat{\psi}_1(s)^r}{(s+\lambda)^{r+1}}\right]^{k-i-j} \qquad (8.19)$$

The inversion of the above equation gives an explicit expression for $P_{1,0}(t)$ as follows,

$$P_{0,1}(t) = \sum_{k=0}^{\infty} \sum_{i=0}^{k} \sum_{j=0}^{k-i} (-1)^{i+j} \gamma_1^k \gamma_2^{k-i-j} \binom{k}{i}\binom{k-i}{j}\left(\frac{\xi_1}{\gamma_1}\right)^i e^{-\left(\lambda+\gamma_1\right)t} \frac{t^k}{k!} * \phi_1(t)^i * \phi_m(t)^{k-i}$$

$$* \psi_m(t)^{k-i-j} * \left[\mu \sum_{m=1}^{\infty} \beta^{1-m}(I_{m-1}(\alpha t) - I_{m+1}(\alpha t))e^{-(\lambda+\mu)t}\right]^{*(k-i)} \qquad (8.20)$$

$$* \left[\sum_{r=0}^{\infty} \xi_2^r e^{-\lambda t}\frac{t^r}{r!} * \psi_1(t)^r\right]$$

where '$*$' denotes convolution while '$* (k - i)$' and '$* (k - i - j)$' represent $(k - i)-$ fold convolution and $(k - i - j)$-fold convolution, respectively.

Thus, equations (8.16), (8.10), (8.11), (8.17), and (8.20) represent all system size probabilities $P_{i,n}(t)$, $i = 0, 1, 2$, $n = 0, 1, 2, \dots$ in terms of modified Bessel function of the first kind.

8.4 TIME DEPENDENT MEAN AND VARIANCE

In this section, time-dependent expected system size and variance are derived.

8.4.1 Mean

Let $X(t)$ denote the number of jobs in the system at time t. The average number of jobs in the system at time t is given by

$$m(t) = E(X(t)) = \sum_{n=1}^{\infty} n\Big(P_{0,n}(t) + P_{1,n}(t) + P_{2,n}(t)\Big)$$

$$m(0) = \sum_{n=1}^{\infty} n\Big(P_{0,n}(0) + P_{1,n}(0) + P_{2,n}(0)\Big) = 0$$

$$m'(t) = \sum_{n=1}^{\infty} n\Big(P'_{0,n}(t) + P'_{1,n}(t) + P'_{2,n}(t)\Big)$$

Using equations (8.1), (8.2), (8.4) and (8.6) and after some algebra, we have the following equation

$$m'(t) = \lambda - \mu \sum_{n=1}^{\infty} P_{0,n}(t) - \xi \sum_{n=1}^{\infty} n\Big(P_{1,n}(t) + P_{2,n}(t)\Big)$$

Integrating the above equation w.r.t. t and applying the initial condition $m(0) = 0$, the solution of the above equation can be obtained as follows;

$$m(t) = \lambda t - \mu \sum_{n=1}^{\infty} \int_0^t P_{0,n}(u)\,du - \xi \sum_{n=1}^{\infty} n\left[\int_0^t P_{1,n}(u)\,du + \int_0^t P_{2,n}(u)\,du\right] \quad (8.21)$$

where $P_{0,n}(t)$, $P_{1,n}(t)$ and $P_{2,n}(t)$ are given by equations (8.16), (8.10) and (8.11), respectively.

8.4.2 Variance

Let $X(t)$ denote the number of jobs in the system at time t. The variance of jobs in the system at time t is given by

$$Var(X(t)) = E(X^2(t)) - [E(x)]^2$$
$$Var(X(t)) = k(t) - [m(t)]^2$$

Where

$$k(t) = E(X^2(t)) = \sum_{n=1}^{\infty} n^2\Big(P_{0,n}(t) + P_{1,n}(t) + P_{2,n}(t)\Big)$$

Also,

$$k(0) = \sum_{n=1}^{\infty} n^2\Big(P_{0,n}(0) + P_{1,n}(0) + P_{2,n}(0)\Big) = 0$$

And

$$k'(t) = \sum_{n=1}^{\infty} n^2\left(P'_{0,n}(t) + P'_{1,n}(t) + P'_{2,n}(t)\right)$$

Using equations (8.1), (8.2), (8.4) and (8.6) and after some algebra, we will have

$$k'(t) = 2\lambda m(t) + \lambda - \mu \sum_{n=1}^{\infty}(2n-1)P_{0,n}(t) - \xi \sum_{n=1}^{\infty} n(2n-1)P_{1,n}(t)$$

$$- \xi \sum_{n=1}^{\infty} n(2n-1)P_{2,n}(t)$$

Integrating the above equation w.r.t. t and applying the initial condition $k(0) = 0$, we can
obtain

$$k(t) = 2\lambda \int_0^t m(u)\,du + \lambda t - \mu \sum_{n=1}^{\infty}(2n-1)\int_0^t P_{0,n}(u)\,du$$

$$- \xi \sum_{n=1}^{\infty} n(2n-1)\int_0^t P_{1,n}(u)\,du$$

$$- \xi \sum_{n=1}^{\infty} n(2n-1)\int_0^t P_{2,n}(u)\,du$$

Substituting the above equation in equation (4.2), we will have

$$Var(X(t)) = 2\lambda \int_0^t m(u)\,du + \lambda t - \mu \sum_{n=1}^{\infty}(2n-1)\int_0^t P_{0,n}(u)\,du$$

$$- \xi \sum_{n=1}^{\infty} n(2n-1)\int_0^t P_{1,n}(u)\,du$$

$$- \xi \sum_{n=1}^{\infty} n(2n-1)\int_0^t P_{2,n}(u)\,du + [m(t)]^2$$

where $P_{0,n}(t)$, $P_{1,n}(t)$, $P_{2,n}(t)$ and m(t) are given by equations (8.16), (8.10), (8.11) and (8.21), respectively.

8.5 SPECIAL CASE

When $\gamma_1 = \gamma$ and $\gamma_2 = 0$, this system becomes an $M/M/1$ queueing system with impatient customers and multiple vacations. Then, from equation (8.15), we will have

$$P_{0,n}(t) = \gamma \int_0^t \sum_{m=1}^{\infty} P_{1,m}(u)\beta^{n-m}[I_{n-m}(\alpha(t-u))$$

$$- I_{n+m}(\alpha(t-u))]e^{-(\lambda+\mu)(t-u)}du \qquad (8.22)$$

Using equation (8.19) and after some algebra, we can derive

$$\hat{P}_{1,0}(s) = \sum_{k=0}^{\infty} \sum_{i=0}^{k} (-1)^i \gamma^k \binom{k}{i}\left(\frac{\xi}{\gamma}\right)^i \frac{\hat{\phi}_1(s)^i}{(s+\lambda)^{r+1}}\left[\sum_{m=1}^{\infty}\left(\frac{p-\sqrt{p^2-\alpha^2}}{\alpha\beta}\right)^m \hat{\phi}_m(s)\right]^{k-i}$$

Taking Laplace transform of the above equation, we have

$$P_{1,0}(t) = \sum_{k=0}^{\infty} \sum_{i=0}^{k} (-1)^i \gamma^k \binom{k}{i}\left(\frac{\xi}{\gamma}\right)^i e^{-\lambda t}\frac{t^k}{k!} * \phi_1(t)^i$$

$$* \left[\mu \sum_{m=1}^{\infty} \beta^{1-m}[I_{m-1}(\alpha t) - I_{m+1}(\alpha t)]e^{-(\lambda+\mu)*t} * \phi_m(t)\right]^{*(k-i)}$$

equation (8.22) and (5.2) agree with equations (4.5) and (4.13) of [23].

8.6 NUMERICAL ILLUSTRATIONS

Even though we have the explicit expressions for the transient state probabilities, it is important to give a visualization for the solutions in practical situations. Therefore, in this section, some of the numerical examples are illustrated to present the behavior of the transient probabilities of the system, when the server is in a busy period and vacation states against time. Furthermore, numerical examples are extended to show the variations of the time dependent mean and variance of the system size against time t. Although, this system has infinite capacity, capacity of the system is restricted to 20 considering the purpose of numerical solutions.

Figures 8.1 and 8.2 are plotted to explain the behavior of $P_{0,n}(t)$ against time t for varying values of n with parameters $\mu = 3.9$, $\gamma_1 = 0.03$, $\gamma_2 = 0.03$, $\xi = 0.01$ and $\lambda = 2.4$. It can be seen that all the values for $P_{0,n}(t)$ start at 0 and finally, tend to settle in a steady state as time progresses.

Figures 8.3 and 8.4 also are used to plot the graph of $P_{1,n}(t)$ against time t for varying values of n with the same parameter values. Here also, all the values for $P_{1,n}(t)$ start at 0 and reach the steady state except $P_{1,0}(t)$.

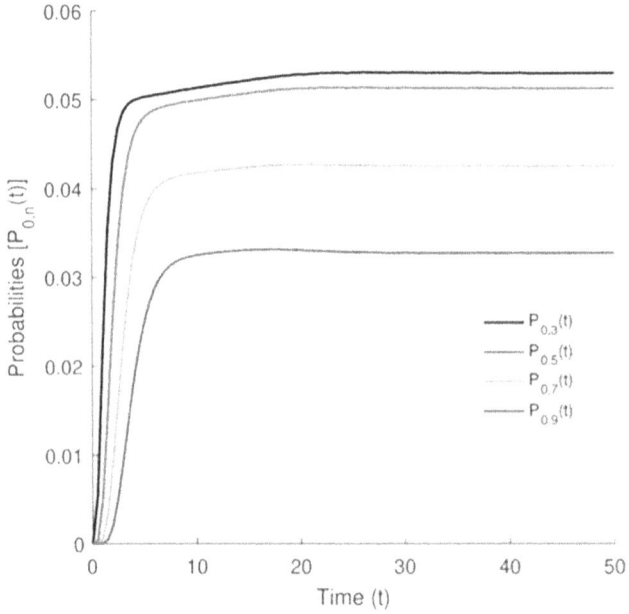

Figure 8.1 Behavior of $P_{0,n}(t)$ against t for varying values of n.

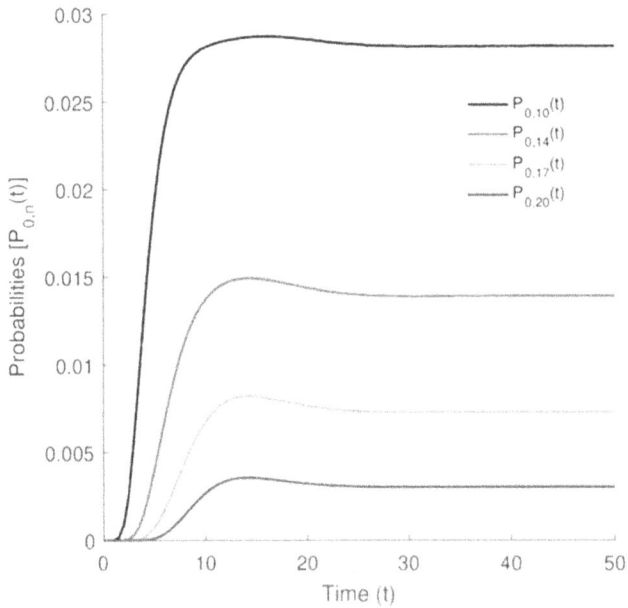

Figure 8.2 Behavior of $P_{0,n}(t)$ against t for varying values of n.

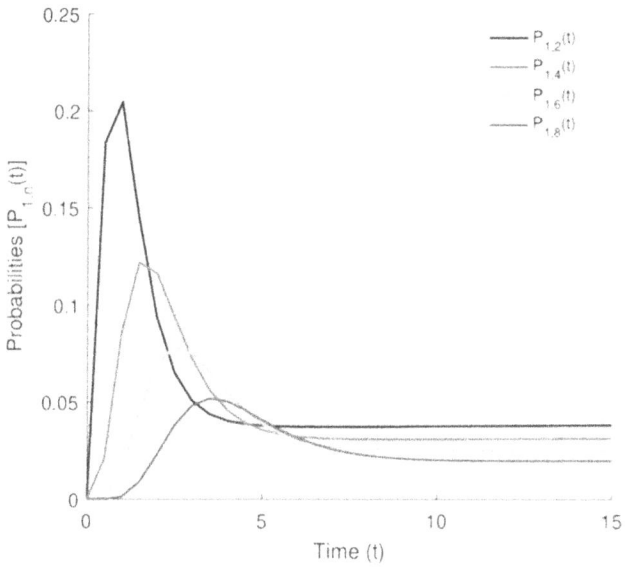

Figure 8.3 Behavior of $P_{1,n}(t)$ against t for varying values of n.

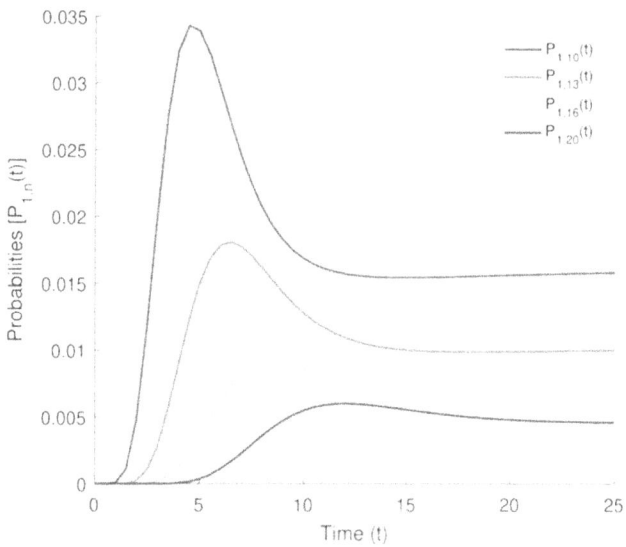

Figure 8.4 Behavior of $P_{1,n}(t)$ against t for varying values of n.

Figures 8.5 and 8.6 show the behavior of $P_{2,n}(t)$ against time t for varying values of n with same parameter values. The values of $P_{2,n}(t)$ start at 0 as previous cases and become steady state with time t.

Figure 8.7 illustrates the behavior of $P_{1,0}(t)$ against time t for varying arrival rates with same parameter values. It can be seen that possibility of having zero customers in vacation type-1 discreases with the increase of the arrival rate. Figure 8.8 depicts the same behavior of $P_{2,0}(t)$ against time t with same parameter values.

Figure 8.9 illustrates the variation of the expected number of customers in the queue against time t for $\mu = 3.9$, $\gamma_1 = 0.03$, $\gamma_2 = 0.05$, $\xi = 0.01$ and varying values of λ (2, 2.5, 3, 3.5). It is clear that expected number of customers in the queue takes higher values with the increase of λ. Figure 8.10 depicts the behavior of the variance of the system size against time t, where with increase of λ, the variance of the number of customers in the system increases.

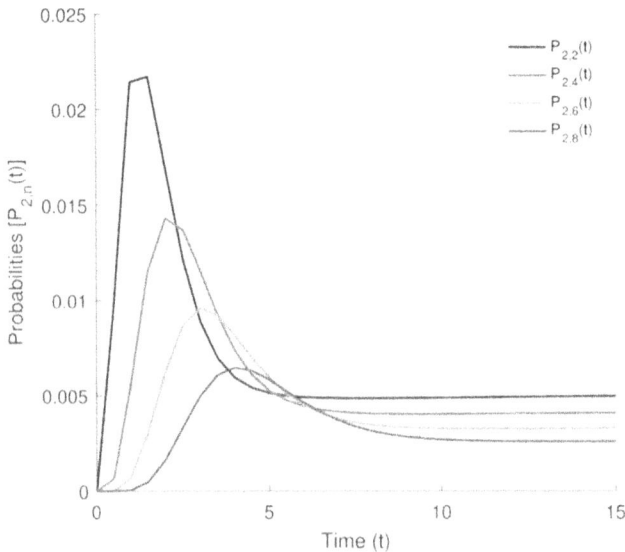

Figure 8.5 Behavior of $P_{2,n}(t)$ against t for varying values of n.

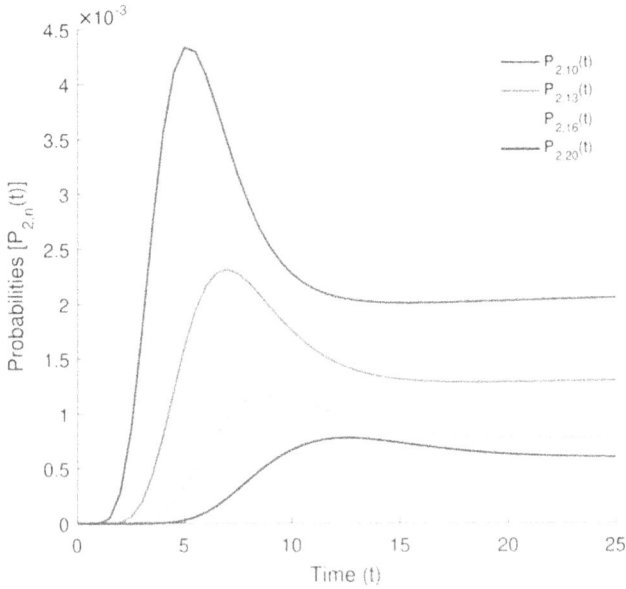

Figure 8.6 Behavior of $P_{2,n}(t)$ against t for varying values of n.

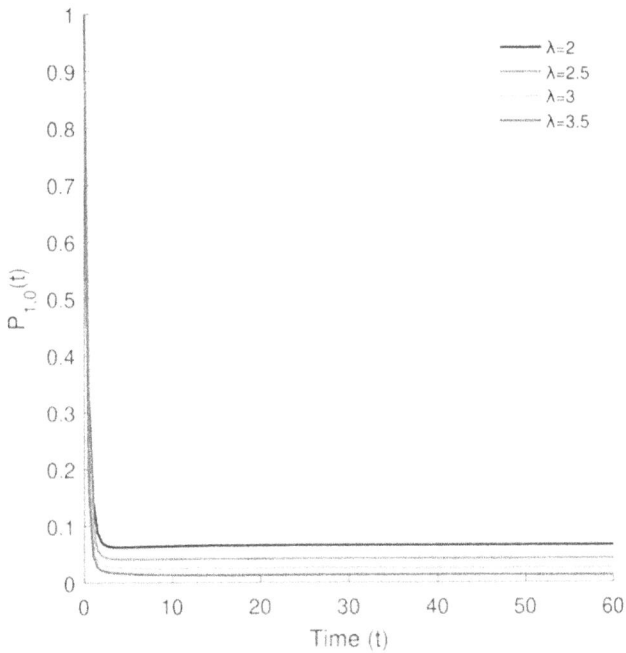

Figure 8.7 Behavior of $P_{1,0}(t)$ against t for varying values of λ.

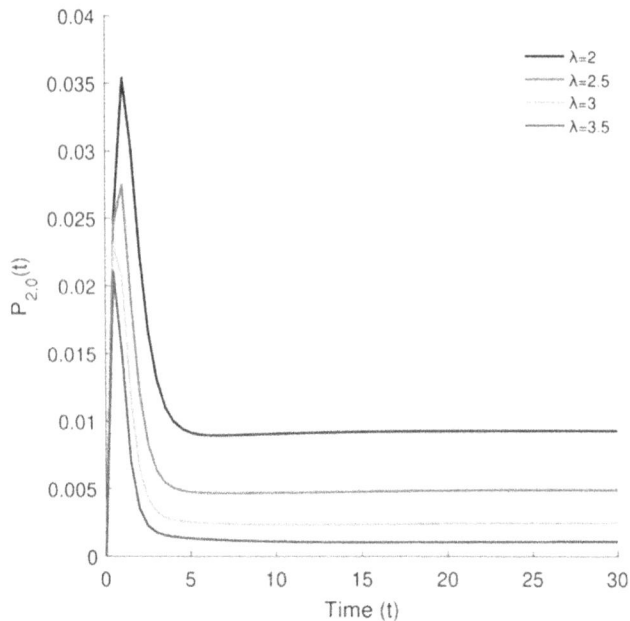

Figure 8.8 Behavior of $P_{2,0}(t)$ against t for varying values of λ.

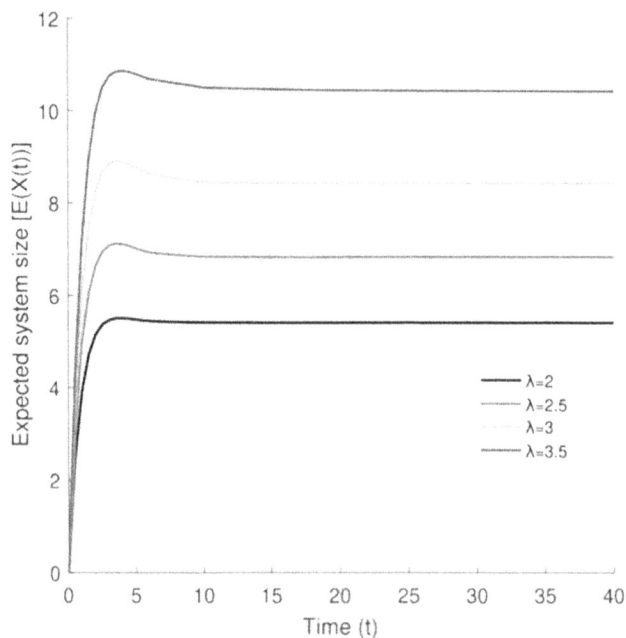

Figure 8.9 Behavior of mean against t for varying values of λ.

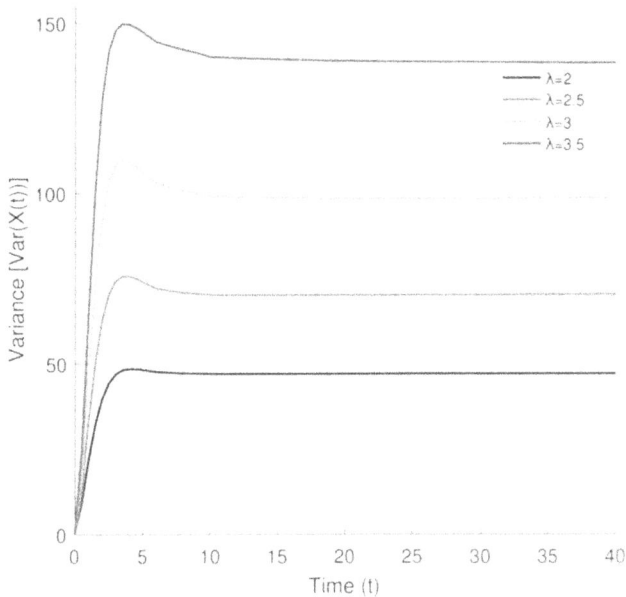

Figure 8.10 Behavior of variance against *t* for varying values of λ.

8.7 CONCLUSIONS

A single server queueing model subjected to differentiated vacations with customer impatience is analyzed in transient regime. The explicit solution is obtained in terms of the modified Bessel function of first kind. Additionally, the mean and variance t are obtained as the performance measures. A numerical example is presented to study the behavior of the system. It is observed that with the progression of time the time-dependent probabilities converge to the corresponding steady-state probabilities.

REFERENCES

1. Levy Y. and Yechiali U., Utilization of idle time in M/G/1 queueing system, Management Science, 22(2) (1975) 202–211.
2. Tian N. and Zhang Z., Vacation Queueing Models-Theory and Applications, (2006) Springer-Verlag, New York.
3. Doshi B., Queueing systems with vacations-A survey, Queueing Systems, 1 (1986) 29–66.
4. Takagi H. 1991. Queueing Analysis: A Foundation of Performance Evaluation, 1, (1991) North-Holland, Amsterdam.
5. Servi L. D. and Finn S. G., M/M/1 queues with working vacations (M/M/1/WV), Performance Evaluation, 50(1) (2002) 41–52.

6. Wu D. and Takagi H., M/G/1 queue with multiple working vacations, Quality Technology and Performance Evaluation, 63(7) (2006) 654–681.
7. Baba Y., Analysis of a G/M/1 queue with multiple working vacations, Performance Evaluation, 63(2) (2005) 201–209.
8. Arumuganathan R. and Jeyakumar S., Steady state analysis of a bulk queue with multiple vacations, setup times with N-policy and close down times, Applied Mathematical Modelling, 29(10) (2005) 972–986.
9. Kalidass K., Gnanaraj J., Gopinath S. and Ramanath K., Transient analysis of an M/M/1 queue with a repairable server and multiple vacations, International Journal of Mathematics in Operational Research, 6(2) (2014) 193–216.
10. Kalidass K. and Ramanath K., Time dependent analysis of M/M/1 queue with server vacations and a waiting server. In: The 6th International Conference on Queueing Theory and Network Applications, QTNA'11, Seoul, Korea. (2011) pp.77–83.
11. Indra and Renu, Transient analysis of Markovian queueing model with Bernoulli schedule and multiple working vacation, International Journal of Computer Applications, 20(5) (2011) 43–48.
12. Sudesh R. and Raj L. F., Computational analysis of stationary and transient distribution of single server queue with working vacation, In: Krishna P. V., Babu M. R., Ariwa E. (eds) Global Trends in Computing and Communication Systems. Communications in Computer and Information Science 269, (2012) Springer, Berlin, Heidelberg.
13. Yang D. Y. and Wu Y. Y., Transient behavior analysis of a finite capacity queue with working break-downs and server vacations, In: Proceedings of the International Multi-Conference of Engineers and Computer Scientists 2. Hong Kong (2014).
14. Kalidass K. and Ramanath K., Transient analysis of an M/M/1 queue with multiple vacations. Pakistan Journal of Statistics and Operation Research, 10(1) (2014) 121–130.
15. Vijayashree K. V. and Janani B., Transient analysis of an M/M/1 queue with multiple exponential vacation and N-policy. Pakistan Journal of Statistics and Operation Research, 11(4) (2015) 587–600.
16. Gahlawat V. K., Indra, Rahul K., Mor R. S. and Malik M., A two state time-dependent bulk queue model with intermittently available server, Reliability: Theory and Applications, 16 (2021) 114–122.
17. Vijayashree K. V. and Janani B., Transient analysis of an M/M/1 queueing system subject to differentiated vacations, Quality Technology and Quantitative Management, 15(6) (2018) 730–748.
18. Ibe O. C. and Isijola O. A., M/M/1 multiple vacation queueing systems with differentiated vacations, Modelling and Simulation in Engineering, 6 (2014) 1–16.
19. Altman E. and Yechiali U., Analysis of customers' impatience in queues with server vacation, Queueing Systems, 52(4) (2006) 261–279.
20. Altman E. and Yechiali U., 2008. Infinite server queues with systems' additional task and impatient customers, Probability in the Engineering and Informational Sciences, 22(4) (2008) 477–493.
21. Perel N. and Yechiali U., Queues with slow servers and impatient customers, European Journal of Operational Research, 201(1) (2010) 247–258.

22. Yue D., Yue W. and Xu G., Analysis of customers' impatience in an M/M/1 queue with working vacations. Journal of Industrial and Management Optimization, 8(4) (2012) 895–908.
23. Ammar S. I., Transient analysis of an M/M/1 queue with impatient behavior and multiple vacations, Applied Mathematics and Computation, 260 (2015) 97–105.
24. Ammar S. I., Transient solution of an M/M/1 vacation queue with a waiting server and impatient customers, Journal of the Egyptian Mathematical Society, 25 (2017) 337–342.
25. Gradshteyn I., Ryzhik I., Jeffery A. and Zwillinger D., Table of Integrals, Series, and Products, 7, (2007) Academic Press, Elsevier.
26. Lorentzen L. and Waadeland H., Continued Fractions with Applications, 3, (1992) Elsevier, Amsterdam.

APPENDIX A EXPRESSION FOR $\phi_n(t)$ AND $\psi_n(t)$

We know that

$$\hat{\phi}_n(s) = \left(\frac{\lambda}{\xi_1}\right)^n \frac{1}{\prod_{i=1}^n \left(\frac{s+\gamma_1}{\xi_1} + i\right)} \frac{{}_1F_1\left(n+1; \frac{s+\gamma_1}{\xi_1} + n + 1; -\frac{\lambda}{\xi_1}\right)}{{}_1F_1\left(n; \frac{s+\gamma_1}{\xi_1} + n; -\frac{\lambda}{\xi_1}\right)} \qquad (A.1)$$

Using the definition of confluent hyper-geometric function, we will have

$${}_1F_1\left(n+1; \frac{s+\gamma_1}{\xi_1} + n + 1; -\frac{\lambda}{\xi_1}\right) = \sum_{k=0}^{\infty} \frac{(n+1)_{(k)}}{\left(\frac{s+\gamma_1}{\xi_1} + n + 1\right)_{(k)}} \frac{\left(-\frac{\lambda}{\xi_1}\right)^k}{k!}$$

Since

$$\frac{(n+1)_{(k)}}{k!} = \binom{n+k}{k}$$

and

$$\left(\frac{s+\gamma_1}{\xi_1} + n + 1\right)_{(k)} = \left(\frac{s+\gamma_1}{\xi_1} + n + 1\right)\left(\frac{s+\gamma_1}{\xi_1} + n + 2\right)\ldots\ldots\ldots$$

$$\left(\frac{s+\gamma_1}{\xi_1} + n + k\right)$$

$${}_1F_1\left(n+1; \frac{s+\gamma_1}{\xi_1} + n + 1; -\frac{\lambda}{\xi_1}\right) = \sum_{k=0}^{\infty} \frac{\binom{n+k}{k}}{\prod_{i=n+1}^{n+k}\left(\frac{s+\gamma_1}{\xi_1} + i\right)}\left(-\frac{\lambda}{\xi_1}\right)^k$$

Then

$$\frac{{}_1F_1\left(n+1; \frac{s+\gamma_1}{\xi_1} + n + 1; -\frac{\lambda}{\xi_1}\right)}{\prod_{i=1}^n\left(\frac{s+\gamma_1}{\xi_1} + i\right)} = \xi_1^n \sum_{k=0}^{\infty} \frac{\binom{n+k}{k}}{\prod_{i=1}^{n+k}\left(s + \gamma_1 + i\xi_1\right)}(-\lambda)^k$$

Taking the partial fractions, we will have

$$
\frac{{}_1F_1\left(n + 1; \frac{s+\gamma_1}{\xi_1} + n + 1; -\frac{\lambda}{\xi_1}\right)}{\prod_{i=1}^{n}\left(\frac{s+\gamma_1}{\xi_1} + i\right)}
$$

$$
= \xi_1^n \sum_{k=0}^{\infty} \binom{n+k}{k} \frac{(-\lambda)^k}{\xi_1^{n+k-1}} \sum_{i=1}^{n+k} \frac{(-1)^{i-1}}{(i-1)!(n+k-i)!} \frac{1}{(s+\gamma_1+i\xi_1)} \tag{A.2}
$$

when $n = 0$,

$$
{}_1F_1\left(1; \frac{s+\gamma_1}{\xi_1} + 1; -\frac{\lambda}{\xi_1}\right) = \sum_{k=0}^{\infty} \frac{(-\lambda)^k}{\prod_{i=1}^{k}(s+\gamma_1+i\xi_1)} = \sum_{k=0}^{\infty} (-\lambda)^k a_k(s)
$$

Where

$$
\begin{aligned}
a_k(s) &= \frac{1}{\prod_{i=1}^{k}(s+\gamma_1+i\xi_1)} \\
&= \frac{1}{\xi_1^{k-1}} \sum_{r=1}^{k} \frac{(-1)^{r-1}}{(r-1)!(k-r)!} \frac{1}{(s+\gamma_1+r\xi_1)}, \quad k = 1, 2, 3, \ldots
\end{aligned}
$$

and $a_0(s) = 1$.

Using the identity given in [25]

$$
\left[{}_1F_1\left(1; \frac{s+\gamma_1}{\xi_1} + 1; -\frac{\lambda}{\xi_1}\right)\right]^{-1} = \sum_{k=0}^{\infty} b_k(s)\lambda^k \tag{A.3}
$$

where $b_0(s) = 1$ and for $k = 1, 2, 3, \ldots$

$$
b_k(s) = \begin{vmatrix}
a_1(s) & 1 & & & & \\
a_2(s) & a_1(s) & & & & \\
a_3(s) & a_2(s) & a_1(s) & \cdots & & \\
\cdots & \cdots & \cdots & \cdots & \cdots & \\
a_{k-1}(s) & a_{k-2}(s) & a_{k-3}(s) & \cdots & a_1(s) & 1 \\
a_k(s) & a_{k-1}(s) & a_{k-2}(s) & \cdots & a_2(s) & a_1(s)
\end{vmatrix}
$$

$$b_k(s) = \sum_{i=1}^{k} (-1)^{i-1} a_i(s) b_{k-i}(s)$$

By substituting the equations (A.2) and (A.3) to the equation (5.2), we can obtain

$$\hat{\phi}_n(s) = \lambda^n \sum_{j=0}^{\infty} (-\lambda)^j \binom{n+j}{j} a_{n+j}(s) \sum_{k=1}^{\infty} b_k(s) \lambda^k$$

Taking inverse Laplace transform of the above equation, we have

$$\phi_n(t) = \lambda^n \sum_{j=0}^{\infty} (-\lambda)^j \binom{n+j}{j} a_{n+j}(t) * \sum_{k=1}^{\infty} b_k(t) \lambda^k$$

where

$$a_k(t) = \frac{1}{\xi_1^{k-1}} \sum_{r=1}^{k} \frac{(-1)^{r-1}}{(r-1)!(k-r)!} e^{-(\gamma_1 + r\xi_1)t}, \quad k = 1, 2, 3, \ldots$$

$$b_k(t) = \sum_{i=1}^{k} (-1)^{i-1} a_i(t) * b_{k-i}(t), \quad k = 2, 3, 4, \ldots; \ b_1(t) = a_1(t)$$

By using the same procedure which has been used to evaluate $\phi_n(t)$, we can find $\psi_n(t)$ as follows,

$$\psi_n(t) = \lambda^n \sum_{i=0}^{\infty} (-\lambda)^i \binom{n+i}{i} d_{n+i}(t) * \sum_{m=1}^{\infty} g_m(t) \lambda^m$$

where

$$d_k(t) = \frac{1}{\xi_2^{k-1}} \sum_{r=1}^{k} \frac{(-1)^{r-1}}{(r-1)!(k-r)!} e^{-(\gamma_2 + r\xi_2)t}, \quad k = 1, 2, 3, \ldots$$

$$g_k(t) = \sum_{i=1}^{k} (-1)^{i-1} d_i(t) * g_{k-i}(t), \quad k = 2, 3, 4, \ldots; \ g_1(t) = d_1(t)$$

B CONFLUENT HYPER-GEOMETRIC FUNCTION

In this section, the definition of confluent hyper-geometric function and some properties of this function are expressed.

The confluent hyper-geometric function is denoted by $_1F_1(a; c; z)$ and is defined by the power

series

$$
\begin{aligned}
_1F_1(a; c; z) &= 1 + \frac{a}{c}\frac{z}{1!} + \frac{a(a+1)}{c(c+1)}\frac{z^2}{2!} + \dots \\
&= \sum_{k=0}^{\infty} \frac{(a)_{(k)}}{(c)_{(k)}}\frac{z^k}{k!}
\end{aligned}
\tag{B.1}
$$

provided that $_1F_1(a; c; z)$ does not exist when c is a negative integer. Here $(a)_{(k)}$ is the rising factorial function (the Pochhammer symbol) which is defined as

$$
\begin{aligned}
(a)_{(0)} &= 1 \\
(a)_{(n)} &= a(a+1)(a+2)\dots\dots\dots(a+n-1)
\end{aligned}
$$

the Pochhammer symbol has the following characteristic

$$
\frac{(a)_{(k)}}{k!} = \binom{a+k}{k}
$$

It is observed that

$$
_1F_1(0; c; z) = 1
$$

The recurrence relation for the confluent hyper-geometric function is given by

$$
c(c-1)_1F_1(a-1; c-1; z) - az_1F_1(a+1; c+1; z)
$$
$$
= c(c-1-z)_1F_1(a; c; z)
$$

The quotient of two hyper-geometric functions may be expressed as continued fractions. The following identity was developed by Lorentzen and Waadeland [26].

$$
\frac{_1F_1(a+1; c+1; z)}{_1F_1(a; c; z)} = \frac{c}{c-z} \frac{(a+1)z}{+c-z+1} \frac{(a+2)z}{+c-z+2} \dots \dots
$$

which can be rewritten as

$$c\frac{{}_1F_1(a;\,c;\,z)}{{}_1F_1(a+1;\,c+1;\,z)} - (c - Z) = \frac{(a+1)z}{c-z+1}\frac{(a+c)z}{+c-z+2}\ldots\ldots \qquad (B.2)$$

and

$$\sum_{k=0}^{\infty}\frac{(a)_{(k)}}{(c)_{(k)}}\frac{y^k}{k!}{}_1F_1(a+k;\,c+k;\,x) = {}_1F_1(a;\,c;\,x+y) \qquad (B.3)$$

Chapter 9

Application of Error Correction Model (ECM) in stabilizing/adjusting fiscal burden post COVID situation

Samir Ul Hassan and Biswambhar Mishra

9.1 INTRODUCTION

When government expenses surpass its revenue collection, it has a budget deficit. When a government incurs a deficit, it can meet this deficit by the following means (a) it can run down its cash reserves, (b) It can sell its assets like properties, (c) It can print more currency and use it (d) It can borrow and spend [1–3]. Note that the second method of meeting the deficit does not increase the government's indebtedness, though a government seldom adopts this approach. The first and third methods increase the government's currency supply in the market, while the fourth increases the outstanding public debts. Each of these methods has consequences for different sectors of the economy. The government usually combines these methods and uses several alternatives at the same time that will be more beneficial to the country at a particular time. Governments can raise more funds by raising taxes, printing money, domestic or external borrowing, and using previous budget surplus. When a government chooses to borrow instead of setting up additional tax measures to fund its budget deficit, it creates a liability on itself known as "public debt" or "national debt."

Regarding the relationship between external debt and economic growth, a reasonable level of borrowing is likely to enable economic growth through capital accumulation and productivity growth [4]. As the government raises much of its revenue from the population, public debt is seen as an indirect debt of the taxpayers. Public debt is a significant aspect of economies all over the world. It influences economic growth in the short and long run [5,6]. Its effect on the economy has become a topical issue and debate among scholars worldwide [7]. According to [8], debt is a two-edged sword. In other words, it can improve welfare when used appropriately but can also be devastating when used carelessly. This affirmation indicates that borrowing is only suitable under definite circumstances, and the government needs to exercise caution while designing its debt policies. Ogunjimi (2019) opined that if the government undertakes realistic public debts, it will most likely improve economic activities and growth. However, the direction of government spending will determine, to

DOI: 10.1201/9781003356653-9

a large extent, whether the public debt will lead to economic growth. For instance, borrowing to carry out development projects, increasing capital expenditure, and rational investment in productive ventures will lead to economic growth in the long run. Unfortunately, many developing countries borrow for other reasons than as expressed above, which is why their debt profile keeps increasing, investment keeps falling, unemployment rises, and national output falls. The majority of her citizens wallow in poverty.

Over the last two decades, the Indian economy has faced a persistent fiscal deficit and adverse balance of payment problems. This is occasioned by incessant fluctuations in the price of crude oil in the international market since India is spending heavily on crude petroleum imports. These inconsistencies in the price of crude in the international market plunged the country's economy into recession in 2008 and 2011. To stimulate the economy, the government is left with no choice but to engage in borrowing (internal and external). India's public debt stock stood at 70 billion US dollars in December 2019. This represents 70% of the GDP of the country. According to the International Monetary Fund (IMF), India's public debt ratio is projected to jump by 17 percentage points to almost 90% because of increased public spending due to COVID-19. India's public debt ratio has remained stable at about 70% of the Gross Domestic Product (GDP) since 1991. The increase in public spending in response to COVID-19, and the fall in tax revenue and economic activity, will make the public debt ratio jump by 17 percentage points. The ratio is projected to stabilize in 2021 before slowly declining to the end of the projection period in 2025. The pattern of public debt in India is close to the norm around the world. This debt-to-GDP ratio compares a country's public debt to its Gross Domestic Product (GDP). It is often expressed as a percentage. Comparing what a country owes (debt) with what it produces (GDP), the debt-to-GDP ratio reliably indicates a particular country's ability to pay back its debts.

A country with a high debt-to-GDP ratio typically has trouble paying public debts. Assessment of Fiscal situation (relating to taxation, public spending, or public debt).

India has been an important source of growth since the 1991 economic liberalization reforms. Real GDP growth averaged 6.5% between 1991 and 2019, and real GDP per capita was multiplied by four over that period. Real GDP is calculated to evaluate the goods and services at constant prices. Nominal GDP, on the other hand, is the value of GDP at the current prevailing prices. This impressive growth performance helped lift millions of people out of extreme poverty. The extreme poverty rate, measured as the proportion of people whose income is less than $1.90 a day at purchasing power parity (the international poverty line), fell from 45% in 1993 to 13% by 2015. India achieved the millennium development goal of halving poverty by 2015 (from its 1990 level). India has made astonishing progress in other areas. Education enrollment is nearly universal for primary school. Infant mortality rates have been halved since 2000. Access to water, sanitation, electricity, and roads has improved greatly.

To finance human capital development, agriculture, and infrastructure development in the areas of roads, railways, waterways, and power, which will help ensure the growth of the economy the Indian government has continued to borrow funds for these projects. It is believed that with the increase in public debt, the government will invest the borrowed funds on projects that will help the economy's development, reduce the unemployment rate, and generate enough funds to repay the loans. Various research has been carried out on this subject matter, and results show that public debt significantly affects public spending while other results show an insignificant effect. Their findings are contradictory, and it is on this background that this study is intended to fill the knowledge gap on the effects of public debt management on public spending in India. As such, the study seeks to determine if the public debt has helped increase public spending in the country. The study also introduces the debt-to-GDP ratio, which compares a nation's debt to its economic output in the country. In addition to the introduction, section two presented a review of related studies, while section three discussed the methodology on which this study is based. Section four presented the analysis of empirical results, while section five discussed summary and policy recommendations.

9.2 LITERATURE REVIEW

One of the most significant determinants of an economy's growth rate is the rate of investment. Countries with high investments experience high growth rates, while countries with low investments are slow in their growth process Tawiri [9]. An economy grows as her investment grows. The need for more secure economic growth has made developing countries seek to improve their human, institutional and infrastructural capacity. This has often increased government expenditures, and the debt burden is expected to increase with insufficient revenue generation. According to [10], developing countries usually obtain debts because they are in the development stage and need additional support. Therefore, public debt links the saving-investment gap and provides extra investment needed for achieving the desired economic growth. As [11–14] noted, many developing countries have policies to attract foreign capital through loans and other means to improve growth. Also, [15] opined that foreign debt is used to generate continuous economic growth that might have been impossible within the pool of domestic resources and the level of technology available for the country. It is also echoed by [16,17] a that foreign borrowing improved resource availability and contributed to economic growth in South Asia. The rationale behind public debt is discussed in what is known as the debt cycle theory. There are three stages in the debt cycle. In the first stage, countries borrow to create additional resources needed for growth. This enables them to stand on their own feet. By the time they are in the second stage, they continue to borrow because the surplus is perhaps not enough to offset interest payments. In the third stage, they would have generated adequate surplus resources to repay the debts.

Private and public spending are significent factors to determine fiscal burden [18–22]. The difference between private and public spending: private investment means putting your own money at risk in anticipation of realizing a gain later, while public spending means taking and spending someone else's money to support your idea of how you think they should live or to satisfy the special interests that help get you re-elected. The public spending is the key channel through which government development goals can be met which will help grow the economy. Essentially, it involves government spending today to grow the economy [23,24]. However, if not properly managed, public debt could lead to more problems than good. It might result in a debt threshold or public debt overhang, resulting in high interest rates, higher inflation, and crowding out private investment [25–27]. For example, creditors might set higher interest rates due to low confidence in the ability of the country to settle its debts. As a result, higher interest rates stimulate high debt costs, forcing the government to impose an additional tax on the citizens [23], inducing the likelihood of economic depression and lower government expenditure in other areas.

Most importantly, higher interest rates may result in low investment, leading to sluggish economic growth in the rest of the economy. This can concurrently stimulate the current account deficit and decline in economic growth, forcing the country to borrow more and thus increase its debt service [28,29]. Additionally, when the debt is accumulated, the cost of servicing this debt would come from taxes on future production. As a result, the investment would be discouraged, crowding out investment [30]. When used correctly, public debts improve the standard of living in a country. It allows the government to construct new roads and bridges, improve education and job training and provide pensions. This encourages citizens to spend more now instead of saving for retirement. This spending by private citizens further boosts economic growth.

9.3 METHODOLOGY

The study is primarily based on secondary data from 1985–86 to 2017–18. The data used in this study has been collected from the Reserve Bank of India (RBI), the Center for Monitoring India Economy (CMIE), and CAG India. The econometric framework adopted in this study is based on developments in the co-integration and error correction model suggested by Johansen (1995, 1988). By applying VAR techniques to the time-series data, based on the unit root and multivariate co-integration test results, we can approximate a dynamic structure in which all the variables in both models are initially treated as endogenous. Most time-series analyses demonstrate non-stationary characteristics in their mean or trending pattern. Therefore, to determine the suitable method of time-series econometric analysis, a common approach is to identify the form of the trend in the data and whether individual data series contain unit root characteristics. If the data is trending, then some form of de-trending is needed. The most common de-trending practices are differencing and time-trend regressions. Thus, the first step in co-integration modeling is often taken by testing for unit roots to determine whether trending data should be differenced or regressed on deterministic functions of time. Based on the unit root and co-integration results, we identify the error correction models best fitting for the given data characteristics and provide robust results.

In this study, the ECM model has been used to assess the influence of public debt management on public spending in India. Thus, the basic model specified is as follows:

$$y_t = A_1 y_{t-1} + \ldots \ldots + A_i y_{t-i} + \ldots \ldots + A_p y_{t-p} + B x_t + e_{t-1} \qquad (9.1)$$

Where:

y_t is the vector of endogenous variables(i.e, Public spending)

x_t Is the vector of exogenous variables (public debt management variables)

A_i And B are matrices of coefficients of the variables to be estimated

e_t is a residual *vector*

i is the lag length, p is the maximum lag length, and t is the time period

The above equation (9.1) states that the dependent variable, y_t varies about its time-invariant means is entirely determined by the explanatory variables with coefficient A_i and $B;$ and the (infinite) past history of y_t itself, the exogenous variables x_t and the history of independently and identically distributed shocks, $e_{t-1}, \ e_{t-2} \ldots \ldots$

However, according to the Granger representation theorem by Engle-Granger, (1987), if co-integration is established among a vector of variables in the model, then a valid error correction model may be estimated; if not, then VAR is used. Therefore, in this study, the choice of whether to use VAR, VECM, or ARDL for estimations follows the Granger representation theorem; that is, it is based on co-integration results.

9.3.1 Description of variables

9.3.1.1 Dependent variable

Public Spending: Different authors have identified spending differently over the years, as shown in the literature review. More specifically, however, public spending can be defined as a public policy of the government to supply goods and services to the public and to strengthen economic performance. In our study, government expenditure has been defined as the total aggregate spending of the government in a financial year, or, in other words, government expenditure in our study is the total revenue expenditure and capital expenditure.

9.3.1.2 Independent/explanatory variables

Public Debt: Public debt is the total amount, including total liabilities, borrowed by the government to meet its development budget

    Budget Deficit: A budget deficit is the sum of revenue and capital account deficits.

    Debt Servicing: Debt service is the cash required to cover the compulsory repayment of interest and principal on a debt taken by the government for a particular period.

    Public debt as a share of GDP: The debt-to-GDP ratio is a simple way of comparing a nation's economic output (as measured by gross domestic output) to its debt levels. In other words, this ratio tells analysts how much money the country earns yearly and how that compares to the money that country owes.

9.4 ESTIMATION PROCEDURE

9.4.1 Test for stationarity

Non-stationary data leads to spurious regression due to non-constant mean and variance Dimitrova, (2005) If a series is stationary without any differencing, it is said to be I(0) or integrated of order 0. However, if a series is stationary after the first difference is said to be I(1) or integrated of order 1. To this end, the Augnebted Dicky-Fuller (ADF) (1979) test has been adopted to examine the stationary, or otherwise, of the time series data. The lowest value of the Akaike Information Criterion (AIC) has been used in this to decide the optimal lag length in the ADF and PP regression. These lags were used in ADF and PP regression to ensure that the error term is white noise. If all the variables in an equation are in the integral order of I(1) and the resulting residuals are I(0). According to Engle and Granger (2005), it can be declared that there resides a corresponding error-correction mechanism (ECM or *et-1*). The basic models will be transformed accordingly. The regression form ADF test is in the following form:

$$\Delta y_t = \alpha_0 + \alpha_1 y_{t-1} + \sum_{j=1}^{p} \gamma_j \Delta y_{t-j} + \varepsilon_t \qquad (9.2)$$

Where Δ is the first-difference operator, y_t Is the respective variable of expenditure over time, p is lag, α_0 is constant, α_1 *and* γ_j Are parameters and ε_t denotes stochastic error term.

If $\alpha_1 = 0$, then the series is said to have a unit root and is non-stationary. Hence, if the hypothesis, $\alpha_1 = 0$, is not accepted according to the equation. It can be concluded that the time series does not have a unit root and is integrated of order I(0); in other words, it has stationarity properties.

Similarly, the regression form Phillips-Perron (PP) test is in the following form:

$$y_t = \alpha_0 + \alpha_1 y_{t-1} + \alpha_2 (t - T/2) + \mu_t \tag{9.3}$$

Where α_0, α_1, α_2 are the expected least-squares regression coefficients.

9.4.2 Co-integration test

Johansen's co-integration method is used to verify whether any long-run association exists between the given variables (Johansen, 1995). Johansen's co-integration test employs two test statistics to identify the number of co-integrating vectors: the Trace test and the Max Eigenvalue test. The Trace statistics tests the null hypothesis of r co-integrating vectors/equation in the given series against the alternative hypothesis of n co-integrating equations. The trace statistic test is calculated by using the following expression:

$$LR_{tr}(r/n) = -T * \sum_{i=r+1}^{n} \log(1 - \check{Y}_i) \tag{9.4}$$

Where

$\check{Y}$ trace statistic value, n is the number of variables in the system, and r = 0, 1, 2, ... , n–1 co-integrating equations.

Similarly, the null hypothesis for the Max Eigen value is to test r co-integrating equations against the alternative of r+1 co-integrating equations where r = 0, 1, 2, ... , n–1 and n is the number of variables in the system. The test statistic for the Max Eigen value is computed as

$$LR_{max}(r/n + 1) = -T * \log(1 - \check{Y}) \tag{9.5}$$

where

$\check{Y}$ the Max Eigenvalue and T is the sample size.

In case the Max Eigen value statistic and the Trace statistic yield different results, the trace test statistic will be preferred, as Alexander (2001) suggested. However, it must be noted here that the above method of identifying co-integration is feasible when the VECM method is adopted. If our estimation procedure shows the application ARDL method best suited through stationary test, then the bound test will be used to identify co-integrated equations [31,32].

9.5 ESTIMATED MODEL

After the Johansen co-integration test, the next is to fit the suitable time series model. If co-integration is established between the variables, this implies that there exists a long-run relationship between the variables and that variables are integrated in the same order; hence, the VECM is applied to determine the short-run relationships of co-integrated variables. If co-integration is established between the variables, this implies a long-run relationship exists between the variables, but variables are integrated in a different order; hence, the ARDL is applied to determine the short and long-run relationships of co-integrated variables. On the other hand, if there exists no co-integration, then the VECM is transformed into a Vector autoregressive (VAR) model, followed by impulse analysis, variance decomposition tests have been used to determine short run casual links and response of dependent variable toward independent variable with a period of stability. The estimated model is the modification of equation 9.1 after the co-integration test. ARDL model has been used to identify short-run and long-term relationships between public debt management variables and public spending. The estimated models for our study are as follows:

$$
\begin{aligned}
lnDPS_t = {} & \alpha_1 + \beta_1(lnPD_{t-i}) + \beta_2(lnBD_{t-i}) + \beta_3(lnDS_{t-i}) + \beta_4(lnPDGDP_{t-i}) \\
& + \sum_{i=0}^{n} \theta_1 DlnPD_{t-i} + \sum_{i=0}^{n} \theta_2 DlnBD_{t-i} + \sum_{i=0}^{n} \theta_3 DlnDS_{+t-i} \\
& + \sum_{i=0}^{n} \theta_4 DlnPDGDP_{t-i} + \textstyle\prod ECT_{t-1}
\end{aligned}
\tag{9.6}
$$

Where

D is the different levels of the variable; ln is the natural log form of the respective variable, and $\alpha 1$ is the intercept coefficients. Parameters $\beta_1, \beta_2, \beta_3, \beta_4$ are short-run coefficients *and* $\theta_1, \theta_2, \theta_3, \theta_4$ are the long-run coefficients of the equation. The coefficient of ECM in the equations represents $\prod ECT_{t-1}$ Shows the speed of adjustment toward the long-run equilibrium. The coefficient of adjustment should be negative and statistically significant for convergence.

The study uses the Granger causality testing at the end to understand the feedback and direction of effect between the variables.

9.6 DATA ANALYSIS AND INTERPRETATION

The empirical analysis begins with stationary analysis to understand the basic characteristics of time series data and avoid spurious regression estimates. The data has been converted into a natural log so that coefficients of the equation are treated as elasticity and data is normally distributed.

Table 9.1 Estimated results of the ADF Test

Variables	At Level			First Difference			Order of Integration
	Test Statistic	1%	5%	Test Statistic	1%	5%	
PS	0.25	−3.64	−2.95	−5.40	−3.65	−2.95	I(1)
PD	4.13	−3.64	−2.95	−2.25	−3.65	−2.95	I(0)
BD	2.18	−3.64	−2.95	−3.54	−3.65	−2.95	I(1)
DS	6.08	−3.64	−2.95	−2.52	−3.65	−2.95	I(0)
PDGDP	−1.28	−3.64	−2.95	−4.57	−3.65	−2.95	I(1)

Source: Calculated by Author

The stationarity test has been conducted using the ADF test. The result of the ADF and PP test is shown in Table 9.1.

The result in Table 9.1 revealed that PD and DS are stationary at level, but PS, BD, and PDGDP are not stationary at level. Based on this, we differentiate the variables to see their outcome. The result of the ADF test shown in Table 9.1 shows that PUINV, BD, and PDGDP are stationary at 1st difference. This shows that the variables used in the study are integrated in order 1(0) and I(1), which necessitated the use of Autoregressive distribution lag(ARDL) as the method of data analysis.

9.7 ARDL CO-INTEGRATION RELATIONSHIP

The affirmation of the non-stationarity of the data through the unit root test of ADF permits the determination of the co-integration relationship between the dependent and explanatory variables in the models. Subsequently, we carry on with the bounds test as it can estimate variables at the level and of the first order of integration [31].

Table 9.2 presents the cointegration results of the bounds test. The public debt–investment model has five variables. Therefore, there are four

Table 9.2 Estimated results of ARDL bounds test

Null Hypothesis: No long-run relationships exist		
Test Statistic	**Value**	**k**
F-statistic	5.236177	4
Critical Value Bounds		
Significance	**I0 Bound**	**I1 Bound**
10%	2.45	3.52
5%	2.86	4.01
2.5%	3.25	4.49
1%	3.74	5.06

Source: Calculated by Author

independent variables in the model, hence k = 4. The calculated F-statistics is 5.236177, greater than the lower bounds critical value of 3.74 and the upper critical value of 5.06 at a 1% significance level. Therefore, there is cointegration among the variables, meaning in the long run, the variables move in the same direction [31].

9.8 ESTIMATED RESULTS OF ARDL ERROR CORRECTION MODEL

The estimates of the long-run and short-run results of the ARDL model are presented in Table 9.3 in panels A and B. The lag lengths of (2,1,2,1,1) for independent variables are determined by Akaike Information Criterion (AIC). The long-run estimates of the ARDL model with public debt, budget deficit, debt servicing, and public debt as a ratio of gross fiscal deficit as explanatory variables and public spending as a dependent variable are shown in panel A of the table. Panel A reveals that debt management mechanisms through public debt, budget deficit, and the ratio of public debt to GDP are the key determinants of public spending. The long-run impact of public debt has a positive and significant impact on public spending, as

Table 9.3 Estimated results of the ARDL model

Dependent Variable: Public spending (PS); ARDL (2,1,2,1,1)

Panel A: Estimated Long Run Coefficients

Variable	Coefficient	Std. Error	Prob.
PD	0.034930	0.035813	**0.0411****
BD	−0.621486	0.249995	**0.0219****
DS	−0.636547	0.438787	0.1624
PDGDP	−3.687869	2.093718	**0.0002***
C	632.166215	121.232293	0.0000

Panel B: Error Correction representation for the ARDL model(short-run estimates)

D(lnPS(-1))	0.474039	0.225135	**0.0481****
D(lnPS(-2))	0.507735	0.233964	**0.0422**
D(lnPD)	0.026740	0.029863	0.3812
D(lnBD(-1))	0.216881	0.132408	0.1171
D(lnBD(-2))	−0.422958	0.173839	**0.0245****
D(lnDS(-1))	−0.487308	0.362045	0.1934
D(lnPDGDP)	−2.516600	2.918673	0.3988
ECM(-1)	−0.765550	0.183623	**0.0005***
R-Squared	**0.891**	**Psudo R-Squared**	0.884
F-Stat	**0.000**	**Akaike Info Criterion**	−5.330

Source: Calculated by Author; *, ** Significant 1% and 5% level of Significance

expected. This implies that a 1% increase in public debt will lead to a 0.036% increase in public spending in the long run. This positive long-run relationship is consistent with the finding of [13,33] . The argument for the positive relationship is that public debt enables nations to borrow to carry out development projects, increase capital expenditure, and rational investment in productive ventures will lead to economic growth in the long run. Unfortunately, many developing countries borrow for other reasons than as expressed above, which is why their debt profile keeps increasing, investment keeps falling, unemployment rises, national output falls, and most citizens wallow in poverty.

The budget deficit shows that a 1% increase in budget deficit leads to reduce public spending by 0.62% in the long run, and the coefficients are significant at a 5% level of significance. This is revealed as the gap between revenue and spending increases and revenue falls short of spending, forcing the government to curtail some spending. This is also true in the FRBM act of India, which fixes the level of the budget deficit for the country to have a smooth financial structure. Therefore, keeping that in mind, when the budget deficit increases more than the threshold limit, it forces governments to reduce spending. The results are in line with [4].

Further, the public debt as a ratio to GDP shows a negative and significant relationship with public spending. Table 9.4 reveals that a 1% increase in public debt as a ratio to GDP tends to reduce public spending by 3.6%. The coefficient is significant at a 1% level of significance. This might be due to the fact that a high public debt ratio to GDP represents the output of the economy, and a higher public debt ratio to GDP negatively affects capital stock accumulation and economic growth viz-a-viz higher distortionary tax rates, inflation, long-term interest rates and most importantly the fiscal policies. These results are in line with Tsoulfidis [18].

In contrast, debt servicing shows a negative but insignificant impact on public spending. Therefore, the above results show that, in the long run, public debt management has serious implications for public spending in India. The mismanagement in raising public debt, budget deficit, and the ratio of public debt to GDP lead to serious implications on fiscal policy, especially in managing public spending efficiently.

The next step is to estimate of short-run dynamic coefficients of the ARDL model. Panel B of Table 9.4 shows the short-run dynamic results of the model. The results are generally consistent with the long-run findings regarding signs

Table 9.4 Estimated results of Breusch-Godfrey serial correlation LM test

Breusch-Godfrey Serial Correlation LM Test			
F-statistic	3.340832	Prob. F(2,18)	0.1584
Obs*R-squared	8.392123	Prob. Chi-Square(2)	0.1151

Source: Calculated by Author

and significance. The table reveals that all the variables are statistically significant in the short run except debt servicing to produce a change in public spending. Still, the time lag impact differs in each variable. The table shows that public spending of previous years (lag 1 and lag 2) have a significant positive impact on next year's public spending. This indicates the incremental budgeting practice exercised in India. The short-run coefficients show that a 1% increase previous year and the previous year leads to an increase in public spending in the current year by 0.47 and 0.50%, respectively, and coefficients are significant at 5% significance. Surprisingly, public debt in the short run shows a positive but insignificant impact. It implies that a 1% increase in public debt last year tends to increase public spending in the current year, but the coefficient is insignificant. The budget deficit in the short run has a negative and significant impact on public spending. Panel B shows that a 1% increase in the budget deficit in lag 2 has a negative but significant impact on public spending. In contrast, the debt servicing and ratio of public debt to GDP have no short-run impact.

The error coefficient of the Error Correction Term (ECM), which is denoted by ECM (–1)) is negative(–0.765) and statistically significant at a 5% level of significance. It reveals the evidence of a fast response to bring equilibrium in public spending when there are short-run shocks. The negative coefficient of the error correction model determines the independent variables' speed of adjustment to long-run equilibrium. The negative coefficient indicates that any shock in public spending that takes place in the short run would be corrected in the long run by 76% through changes in the given independent variables. It shows that any fluctuation caused in previous years, or the short run, will bring equilibrium in the long run at 76%, and it will take at least two years to restore any disequilibrium in public spending. The rule of thumb is that the larger the error correction coefficient (in absolute terms), the faster the variables equilibrate in the long run when shocked. The R square of equation (.89) suggests that the 89% variation in public spending is explained by the variables used in the model.

9.9 DIAGNOSTIC TESTS

Various diagnostic tests have been carried out to test the goodness of fit of the ARDL model. Breusch-Godfrey (LM Test) was carried out to know whether the model has the problem of autocorrelation. Jarque-Bera test has been used for normality and the Cusum Test for the stability of coefficients. The results are given in the table below.

From Table 9.4, the p-value is greater than the chosen significance level of 5%, indicating the absence of autocorrelation in the models. The result of the serial correlation shows that the probability value is 0.1584, which is greater than 0.05, implying that we fail to reject H_0 and reject H_1. We then conclude that there is no serial autocorrelation in the model and that the model is appropriate for the study.

9.10 NORMALITY TEST

The normality test was done using the Jarque-Bera Normality test, which requires a series to be normally distributed; the histogram should be bell-shaped, and the Jarque-Bera statistics would not be significant. This implies that the p-value given at the bottom of the normality test table should be greater than the chosen significance level to accept the Null hypothesis that the series is normally distributed. (Figure 9.1)

The result of the normality test shows that the probability value of 0.500113 is greater than 0.05. Based on this, however, we fail to reject H_0 and reject H_1. We then conclude that the residuals are normally distributed and random. (Figure 9.2)

The CUSUM test is the test used to check stability within the model. The results of the stability test show evidence that the model is stable. This is indicated by a movement of blue lines within the figure's critical lines

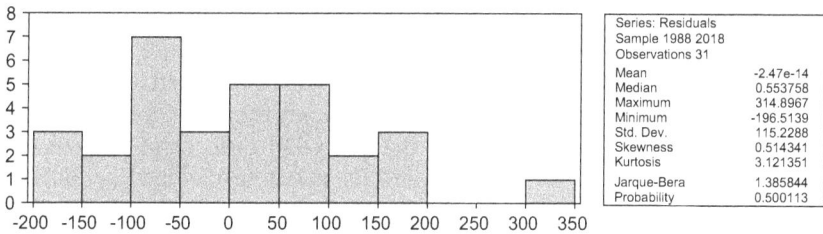

Series: Residuals	
Sample 1988 2018	
Observations 31	
Mean	-2.47e-14
Median	0.553758
Maximum	314.8967
Minimum	-196.5139
Std. Dev.	115.2288
Skewness	0.514341
Kurtosis	3.121351
Jarque-Bera	1.385844
Probability	0.500113

Figure 9.1 Normality text.

Source: Calculated by Author.

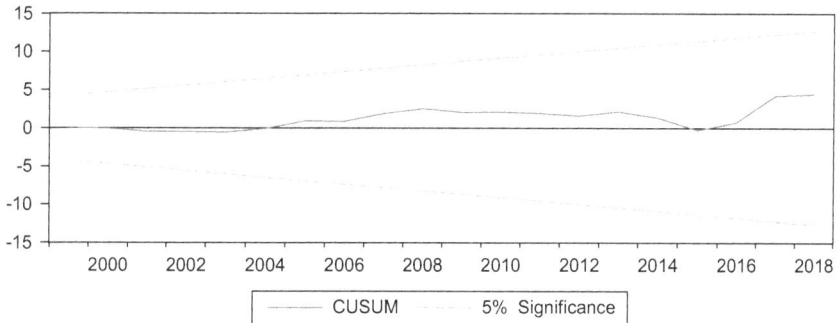

Figure 9.2 Cusum tests of stability.

Source: Calculated by Author.

(two red dotted lines). Therefore, at a 5% significance level, the CUSUM stability test confirms the model's good performance.

9.11 GRANGER CAUSALITY TEST

The study further used the Granger Causality analysis to understand the relationship between public debt management dimensions and public spending. Table 9.5 shows that there exists a unidirectional relationship between public debt and debt servicing on public spending, with causality flowing from public debt and debt servicing to public spending. Also, there is bidirectional causality between budget deficit and public spending.

This shows that a country with low spending will likely have a large government deficit. In contrast, a unidirectional relationship shows that if the borrowed funds are used properly, they will be channeled to productive investments, which will help the economy's growth [34]. From the result, it is confirmed that the nature of poor domestic savings and investment causes higher debt service payments and crowds out available resources for investment in economic and social sectors. The non-causality of debt to GDP ratio also revealed that India had not invested the borrowed funds in projects or investments that can generate enough funds for the country to repay the borrowed fund, since the debt to GDP ratio compares what a country owes with what it produces and its ability to pay back its debts.

Table 9.5 Pairwise Granger causality test on input variables

Null Hypothesis	Obs	F-Statistic	Prob.	Decision
PD does not Granger Cause PUINV	32	3.81910	0.0346	**Reject the Null Hypothesis**
PS does not Granger Cause PD		1.74776	0.1933	Fail to reject the Null Hypothesis
BD does not Granger Cause PUINV	32	4.97064	0.0145	**Reject the Null Hypothesis**
PS does not Granger Cause BD		4.53892	0.0200	**Reject the Null Hypothesis**
DS does not Granger Cause PUINV	32	6.98338	0.0036	**Reject the Null Hypothesis**
PS does not Granger Cause DS		0.65591	0.5270	Fail to reject the Null Hypothesis
PDGDP does not Granger Cause PS	32	1.45659	0.2508	Fail to reject the Null Hypothesis
does not Granger Cause PDGDP		2.91459	0.0714	Fail to reject the Null Hypothesis

Source: Calculated by Author

9.12 SUMMARY AND CONCLUSION

The study has provided conclusive evidence for underlining the fact that public debt has a resounding effect on public spending in India. The ARDL and Granger causality analysis has been used to explore such a relationship. The cointegration test shows the long-run relationship between the investigated variables. The long-run relationship shows that public debt and public spending have a positive relationship. The ECM confirmed that the system could adjust to equilibrium at a speed of 76%. There is a uni-directional Granger causality relationship between public debt and public spending. Therefore, it can be argued that one standard deviation shock in public debt positively affects public spending, with public debt, budget deficit, and the ratio of public debt to GDP having the highest shock impact. The short-run relationship shows that public debt and public spending have positive and insignificant effect in India. The insignificant effect has shown that public debt incurred in India has been used for consumption that does not generate future income rather than on spending. Also, the result shows that in India, huge debt has been used on recurrent expenditures and wasteful projects instead of investing in capital projects or infrastructure that will help increase the tax base and revenue to the government. Equally, the huge debt means that the resources that would have been used for investment are diverted to meeting debt service obligations. The debt servicing and the adjustment policies required to address the debt burden have also worsened investment in such areas as social welfare in the areas of education, health, and communication. In the near term, additional fiscal action should be deployed to support the poor and the vulnerable. This should be accompanied by a credible medium-term fiscal consolidation plan to reinforce market confidence and structural reforms that boost India's growth potential. The effects of COVID-19 will lead to more burdens on health, education, poverty, and nutrition, rendering progress toward the Sustainable Development Goals even more urgent. Macroeconomic and financial stability are important necessary conditions for sustainable development.

REFERENCES

1. Adamu, I. M. (2016). Effects of external debt on public capital investment in Nigeria. Academic Journal of Economic Studies, 2(4), 120–138.
2. Adedoyin, L. I., Babalola, B. M., Otekinri, A. O. & Adeoti, J. O. (2016). External debt and economic growth: Evidence from India. Acta Universitatis Danubius, Economica, 12(6), 179–194.
3. Bhartia, H. L. (2009). Public finance (14th ed.). New Delhi: Vikas Publishing House.

4. Chowdhury, A. R. (2001). External debt, growth, and the HIPC Initiative; Is the country choice too narrow? In Debt relief for poor countries, T. Addison, H. Hansen of F. Torp, eds. Hampshire and New York.

5. Kumar, M. S. & Woo, J. (2015) Public debt and growth. Economica, 5(82), 705–739.

6. Ma'ale, S. A. (2019). The impact of public debt and public spending on economic growth in Jordan. International Journal of Academic Research in Accounting, Finance and Management Sciences, 9(2), 149–157.

7. Herndon, T., Ash, M. & Pollin, R. (2014). Does high public debt consistently stifle economic growth? A critique of Reinhart and Rogoff. Cambridge Journal of Economics, 38(20), 257–279

8. Cecchetti, S., Mohanty, M. S. & Zampolli, F. (2011). The Real Effects of Debt, Bank for International Settlement Working Papers No. 352.

9. Tawiri, N. (2010). *Domestic Investment as a drive of economic growth in Libya*. International Conference on Applied Economics – ICOAE, 759–767.

10. Ajayi, K. (2000). International administration and economic relations in a changing world. Ilorin: Majab Publishers.

11. Mohammad, A. & Sabahat Zareen, F. N.(2000). Export diversification and the structural dynamics in the growth process: The Case of Pakistan. The Pakistan Development Review, Pakistan Institute of Development Economics, 39(4), 573–589.

12. Ncanywa, T. & Masoga, M. M. (2018) Can public debt stimulate public spending and economic growth in South Africa? Cogent Economics & Finance,6(1), 2332-2039.

13. Ogwuma, M. M., Orikara, C. P. & Uruakpa, N. I. (2016). Domestic public debt and public expenditure in India: Any positive correlation on economic growth (1980–2016). A Multidisciplinary Journal of Global Macro Trends, 10(5), 20–45

14. Panizza, U. & Presbitero, A. F. (2012). Public debt and economic growth in advanced economies: A survey. Swiss Journal of Economics and Statistics, 14(9), 175–204.

15. Ahmad, E. (1999). Retiring public debt through privatization. Pakistan Journal of Applied Economics, 15(1), 1–18.

16. Siddiqui, R. & Malik. A. (2002). *"Debt and economic growth in South Asia"*. Paper presented in the 17th Annual General Meeting of PSDE, held in Islamabad in January 2002.

17. Swanson, N. R., & Granger, C. W. J. (1997). Impulse response functions based on a causal approach to residual orthogonalization in vector auto regressions. Journal of American Statistical Association, 92(437), 357–367.

18. Tsoulfidis, L. (2007). Classical economists and public debt. International Review of Economics, 54(1), 1– 12.

19. Udeh, S. N., Ugwu, J. I. & Onwuka, I. O. (2016). External debt and economic growth: The Nigeria Experience. European Journal of Accounting, Auditing, and Financial Research, 4(2), 33–48.

20. United Nations (2009) *The role of public spending in social and economic development*. Public spending: Vital for Growth and Renewal, but should it be a Countercyclical Weapon? New York and Geneva.

21. Wambui, S. K. (2015). *The effects of public debt on private investments and economic growth in Kenya (1980–2013)*. A Research Project Submitted to

the Department of Econometrics and Statistics in the School of Economics In Partial Fulfilment of The Requirements for the Award of Masters of Economics (Econometrics) Degree of Kenyatta University.

22. Richard, S., Ssebulime, K. & Enoch, T. (2018). Uganda's experience with debt and economic growth "an empirical analysis of the effect of public debt on economic growth- 1980-2016" https://ssrn.com/abstract=3497383

23. Hoag, A. J. & Hoag, J. H. (2006). Introductory economics (fourth ed.). Word Scientific Publishing Co PTC Ltd.

24. Ibrahim, M. A. (2016). Public spending in India. does external debt matter?. Academic Journal of Economic Studies, 2(4), 120–138.

25. Boccia, R. (2013). How the United States' high debt will weaken the economy and hurt Americans, backgrounder, The Heritage Foundation. Leadership for America, no. 2768.

26. Broner, F. (2013). Sovereign Debt Markets in Turbulent Times: Creditor Discrimination and Crowding-Out Effects. Unpublished Research Paper, Barcelona Graduate School of Economics.

27. Brooks, C. (2008). Introductory econometrics for finance (2nd ed.). New York: Cambridge University Press.

28. Iyoha, M. A. (1999). An econometric study of debt overhang, debt reduction, investment and economic growth in Nigeria. Ibadan, Nigeria: NCEMA.

29. Kasele, O., Momeka, L., Bahaya, G. & Ntumwa, B. (2019). Impact of public debt on investment: Case of ECGLC. Journal of Finance and Economics, 7(4), 148–163.

30. Tabengwa, G. K. (2014). Impact of shocks to public debt and government expenditure on human capital and growth in developing countries. Journal of Economics and Behavioural Studies, 6(1), 44–67.

31. Pesaran, M. H., Shin, Y. & Smith, R. J. (2001). Bounds testing approaches to the analysis of level relationship. Journal of Applied Econometrics, 16(3), 289–326.

32. Picarelli, M. O., Vanlaer, W. & Marneffe, W. (2019). Does public debt produce a crowding-out effect for public spending in the EU? Working Paper Series 36.

33. Akomolafe, K. J., Bosede, O., Oni, E. & Achukwu, M. (2015). Public debt and private investment in India. American Journal of Economics, 5(5), 501–507.

34. Rajan, R. (2005). Debt relief and growth. Finance and Development, 42(2), 1–4.

Chapter 10

An inventory model with preserving environment for perishable items under learning effect

Mahesh Kumar Jayaswal, Khursheed Alam, Rifaqat Ali, Chandra Shekhar, and Santosh Kumar

10.1 INTRODUCTION

In business, there are several issues between retailer and customer concerning optimal profit along with total cost from one side to the other. Adad and Jaggi [1] have made efforts to solve these type of issues with the help of a model. This model has shown the effects of credit financing policy. Shinn and Hwang [2] estimated that the highest price for the retailer along with lot size operate with the help of a credit financing scheme. In the inventory system, overall cost performs an important role.

Hung and Chung [3] introduced an inventory model which is the enlarged form of Goyal [4] describing how to reduce the total cost with the help of credit financing along with payment rebate scheme. An arithmetic model has been spread for best costing and batch sizing which presumes that buying quantity and selling price both are dissimilar below credit financing policy when demand is a function of selling price. Hung [5] recommended progressing the EOQ model in an easy way and specified new suggestions on how to observe the optimal lot size for the retailer. Luo [6,7] suggested an inventory model for cooperation between retailer and customer below the credit financing scheme. The research work of Goyal et al. [4] has been generalized by Su et al. [8] with the help of credit financing policy and payment rebate. Teng et al. (2006) arranged a two-level credit financing inventory model and in this model the total order is optimized below the permissible delay in amount.

The analysis work of Huang [9] has been explored by Huang [10] who upgraded this model with the help of two-level credit financing as well as restricted storage space. Teng and Goyal [11] introduced a stock model for buyers when the buyer used the credit policy provided by the customer in the industrial sector. Huang [5] stabilized the establishment of total economic order which was increased with the form of Huang [9] with the help of a two-level credit financing scheme.

DOI: 10.1201/9781003356653-10

Huang and Hsu [12] expanded to a two-level credit financing scheme with the help of a partial credit scheme. An economic order quantity formulation was presented by Jaggi et al. [13] introducing an economic order quantity formulation with the help of two-level credit policy with credit dependent demand. Huang's [5] research tasks have been modified by Teng and Chang [14] with the help of a two-level credit financing scheme which is helpful for the consumers. Chen and Kang [15] expanded a model for an industrial system with the help of two-level credit policy with price dependent demand along with satisfaction. Shah et at. [16] provided a large amount of articles for the inventory system with the help of a two-level credit financing scheme. That factor affected the cost of airplanes examined by Wright [17]. The learning effect on seller's ordering policy for defective items with trade credit financing was suggested by Jayaswal [18]. The total manufacturing of profit-making models for items with defective quality subjected to the effect of learning was described by Jaber et al. [19].

10.2 ASSUMPTIONS AND NOTATIONS

10.2.1 Assumptions

- The continuity of replacement is allowed.
- Shortages are not included in this model.
- The time horizon plane has been considered finite.
- The learning effect is involved in holding and ordering cost.
- Demand rate is constant.
- Lots have a constant deterioration rate.
- Lead time is not included in this model.

10.2.2 Notations

D Rate of demand
ξ The cost for preservation
A Set up cost for inventory
P Item cost for selling
θ The cost for deterioration
C The cost for purchasing item
h The cost for holding item
Q Lot size
T Cycle time
$I(t)$ The inventory stock at $0 \leq t \leq T$
$\pi(T)$ Total average cost for buyer

10.3 MATHEMATICAL FORMULATION

Suppose that initially, the inventory level when $t = 0$ is Q. Let $I(t)$ be the inventory level in the time interval $[0, \quad T]$ which reduces due to the demand and deterioration and the total inventory is finished when $t = T$. We can then write it in the form of an ordinary differential equation with boundary condition which is given under below

$$\frac{dI(t)}{dt} + I(t)\theta = -D, \quad \text{when } t \in [0, \quad T] \tag{10.1}$$

Under some circumstance, $I(0) = Q$ and. $I(T) = 0$.

The value of inventory level at time t can be calculated with the help of equation (10.1) using ODE which is given below and taken as a general solution of equation (10.1)

$$I(t) = (De^{\theta(T-t)} - D)\frac{1}{\theta}, \quad \text{when } t \in [0, \quad T] \tag{10.2}$$

Now, we want to find the value of inventory level at the start. Putting $t = 0$ in equation (10.2), we get

$$Q = I(0) = (e^{\theta T}D - D)\frac{1}{\theta} \tag{10.3}$$

The cost of this scenario is given below

$$\text{The set cost per cycle,} \quad A = \left(\frac{C_1}{T} + \frac{C_2}{Tn^\beta}\right) \tag{10.4}$$

$$\text{Holding cost per cycle,} \quad IHC = \frac{\left(h_1 D + \frac{h_2 D}{n^\beta}\right)}{\theta^2 T}(e^{\theta T} - \theta T - 1) \tag{10.5}$$

$$\text{Deterioration cost per cycle,} \quad CQ - CDT = \frac{CDe^{\theta T}}{\theta T} - CD - \frac{CD}{\theta T} \tag{10.6}$$

$$\text{Preservation cost,} \quad PV = \xi\, T \tag{10.7}$$

Now, the total cost for this scenario

$$\pi(T) = \left[\frac{PV}{T} + \frac{CD}{T} + \frac{A}{T} + \frac{IHC}{T}\right] \tag{10.8}$$

10.4 SOLUTION PROCESS

For optimization of cycle length, we have to take $\frac{d\pi(T)}{dT} = 0$ which gives

$$T = T_1(\text{say}) = \sqrt{\frac{2\left(C_1 + \frac{C_2}{n^\beta}\right)}{D\left\{h_1 + \frac{h_2}{n^\beta} + C(\theta)\right\}}} \tag{10.9}$$

Now, we calculate the second derivative

$$\frac{d\pi(T)}{dT} = -\frac{\left(C_1 + \frac{C_2}{n^\beta}\right)}{T^2} + \frac{\left(h_1 + \frac{h_2}{n^\beta}\right)D}{2} + \frac{CD\theta}{2} \text{ and } \frac{d^2\pi(T)}{dT^2} = \frac{2\left(C_1 + \frac{C_2}{n^\beta}\right)}{T^3} \tag{10.10}$$

which gives

$$\text{which gives, } \frac{d^2\pi(T_1)}{dT^2} = \frac{2\left(C_1 + \frac{C_2}{n^\beta}\right)}{T_1^3} > 0 \tag{10.11}$$

Equation (10.11) shows the convexity of total inventory cost, hence the optimal cycle length is

$$T = T_1(\text{say}) = \sqrt{\frac{2\left(C_1 + \frac{C_2}{n^\beta}\right)}{D\left\{h_1 + \frac{h_2}{n^\beta} + C\theta\right\}}} \tag{10.12}$$

10.4.1 Numerical example

$D = 500$ units, $h_1 = 2$, $h_2 = 1$, $C_1 = 30$, $C_2 = 10$, $\beta = 0.23$, $\theta = 0.20$,

$\xi = 0.15$ per items,

$C = \$50$, optimal cycle length, $T^* = 2.2364$ years, n = 5,

and minimum inventory total cost $\pi(T^*) = 3053$ $ per year

10.5 SENSITIVITY ANALYSIS

The sensitivity analysis of this scenario is presented table wise which is given below.

Managerial insights and observation

- From Table 10.1, it is found that the learning rate is increased from 0.23 to 0.27. The retailer's total profit decreases when learning rate is increased and the cycle length is almost fixed. This study informs decision-makers to take account of the learning effect while making decisions which helps them to earn more profit for the organization. Hence the retailer gets more information for the exercise of shipments.
- From Table 10.2, it is found that the number of shipments increases, from 1–5, and the cycle length is increased up to the 5th shipment. The retailer's total cost decreases when the number of shipments increases while cycle length is steadily increased.
- From Table 10.3, it is found that the deterioration rate is increased slowly and the cycle time initially decreases while the seller's overall cost is increased with respect to deterioration rate.
- From Table 10.4, it is found that the preservation cost is increased by up to 0.15 to 3.15 and cycle length is fixed while the seller's overall cost is increased.

Table 10.1 Effect of learning rate on cycle time and total average cost

Learning rate β	Cycle length T (Year)	Retailer's total cost $\pi(T)$ ($)
0.23	2.2364	3053
0.24	2.2364	3047
0.25	2.2365	3040
0.26	2.2364	3032
0.27	2.2364	3028

Table 10.2 Effect of shipment on cycle time and total average cost

Number of shipments (n)	Cycle length T (year)	Retailer's total cost $K(T)$ ($)
1	2.2340	3433
2	2.2345	3347
3	2.2348	3303
4	2.2352	3274
5	2.2365	3053

Table 10.3 Effect of deteriorating rate on cycle time and total average cost

Deterioration rate θ	Cycle time T (Year)	Retailer's total cost $\pi(T)$ ($)
0.10	5.5126	2597
0.15	3.6497	2925
0.20	2.2364	3053
0.25	2.1873	3580
0.30	1.091	3809

Table 10.4 Effect of preservation cost on cycle time and total average cost

Preservation cost ξ/items	Cycle length T (year)	Retailer's total cost ($\pi(T)$ $)
0.15	2.2364	3053
1.15	2.2364	3154
2.15	2.2364	3355
3.15	2.2364	3456

10.6 CONCLUSIONS

This chapter has tried to develop an arithmetic formula to decide cycle length and the correlation with overall cost with the effect of learning put in over the holding cost as well as the ordering cost. The consequences of this model appear to show that the buyer's whole cost comes down as learning rate increases. When items are easily spoiled then preservation ought to be essential to control the deterioration rate; however, the total cost increases. In this chapter, the items disclose that learning notion is very helpful to get less inventory value. The mathematical analysis detected together clearly suggested that the existence of preservation and the effect of learning had a positive effect on the whole cost. The present chapter contributes to more sensible positions such as supply reliance and cloudy environments.

REFERENCES

1. Abad, P. L., & Jaggi, C. K. (2003). A joint approach for setting unit price and the length of the credit period for a seller when end demand is price sensitive. International Journal of Production Economics, 83(2), 115–122.
2. Shinn, S. W., & Hwang, H. (2003). Optimal pricing and ordering policies for retailers under order-size-dependent delay in payments. Computers and Operations Research, 30, 35–50.

3. Huang, Y. F., & Chung, K. J. (2003). Optimal replenishment and payment policies in the EOQ model under cash discount and trade credit. Asia Pacific Journal of Operational Research, 20(2), 177–190.
4. Goyal, S. K. (1985). Economic order quantity under conditions of permissible delay in payments. Journal of the Operational Research Society, 36, 35–38.
5. Huang, Y. F. (2007). Optimal retailer's replenishment decisions in the EPQ model under two levels of trade credit policy. European Journal of Operational Research, 176(3), 1577–1591.
6. Luo, J. (2007). Buyer–vendor inventory coordination with credit period incentives. International Journal of Production Economics, 108(1–2), 143–152.
7. Sarmah, S. P., Acharya, D., & Goyal, S. K. (2007). Coordination and profit sharing between a manufacturer and a buyer with target profit under credit option. European Journal of Operational Research, 182(3), 1469–147
8. Su, C. H., Ouyang, L. Y., Ho, C. H., & Chang, C. T. (2007). Retailer's inventory policy and supplier's delivery policy under two-level trade credit strategy. Asia-Pacific Journal of Operational Research, 24(05), 613–630.
9. Huang, Y. F. (2003). The deterministic inventory models with shortage and defective items derived without derivatives. Journal of Statistics and Management Systems, 6(2), 171–180.
10. Huang, Y. F. (2006). An inventory model under two levels of trade credit and limited storage space derived without derivatives. Applied Mathematical Modelling, 30(5), 418–436.
11. Teng, J. T., & Goyal, S. K. (2007). Optimal ordering policies for a retailer in a supply chain with up-stream and down-stream trade credits. Journal of the Operational Research Society, 58(9), 1252–1255.
12. Huang, Y. F., & Hsu, K. H. (2008). An EOQ model under retailer partial trade credit policy in supply chain. International Journal of Production Economics, 112(2), 655–664.
13. Jaggi, C. K., Goyal, S. K., & Goel, S. K. (2008). Retailer's optimal replenishment decisions with credit-linked demand under permissible delay in payments. European Journal of Operational Research, 190(1), 130–135.
14. Teng, J. T., & Chang, C. T. (2009). Optimal manufacturer's replenishment policies in the EPQ model under two levels of trade credit policy. European Journal of Operational Research, 195(2), 358–363.
15. Chen, L. H., & Kang, F. S. (2010). Integrated inventory models considering the two-level trade credit policy and a price-negotiation scheme. European Journal of Operational Research, 205(1), 47–58.
16. Shah, N. H., Gor, A. S., & Wee, H. M. (2010). An integrated approach for optimal unit price and credit period for deteriorating inventory system when the buyer's demand is price sensitive. American Journal of Mathematical and Management Sciences, 30(3–4), 317–330.
17. Wright, T. P. (1936). Factors affecting the cost of airplanes. Journal of Aeronautical Science, 3, 122–128.
18. Jayaswal, M., Sangal, I., Mittal, M., & Malik, S. (2019). Effects of learning on retailer ordering policy for imperfect quality items with trade credit financing. Uncertain Supply Chain Management, 7, 49–62.
19. Jaber, M. Y., Goyal, S. K., & Imran, M. (2008). Economic Production Quantity model for items with imperfect quality subjected to learning effects. International Journal of Production Economics, 115, 143–150.

Index

ADF, 212
Analysis of Variance, 6
ARDL, 211
Asymptotic Distribution, 4, 11

Bayesian Statistics, 1, 13
Bessel Function, 181, 182, 187,
    190, 199
Budget Deficit, 216
Busy Density Function, 124
Busy Period Analysis, 118

Cloudy Environment, 229
Clustering Algorithm, 158
Coefficient of Determination, 2, 3, 5
Co-Integration, 210, 211
Conventional Bootstrapping, 49
Correlation, 229
Correlation Coefficient, 6
Cosmological Models, 1, 2, 5
Customer Impatience, 182, 199
CUSUm Test, 219
Cycle Time, 225

Debt- to- GDP Ratio, 208
Defective Items, 225
Demand Rate, 225
Deterioration, 225
Deterioration Rate, 228
Differential-Difference Equations, 119
Differentiated Vacations, 182, 183,
    199, 200

ECM, 210
Economic Order Quantity, 225
EM Algorithm, 156
Empirical Mode, 90

External Debt, 207

Fairness Evaluation, 135, 136, 144, 145
First Difference, 212
Fiscal Policy, 217
Fixout, 134, 138, 139
FRBM Act, 217

Gaussian Mixture Model, 155, 156,
    157, 159
Goodness of Fit, 2, 3, 5, 6, 7, 9, 12, 13,
    14, 15, 16, 19, 20

Hard EM, 154, 155

IMF, 208
Improved Sufficient Bootstrapping, 57
Infrastructure Capacity, 209

Jackknife Bootstrap, 49

Lag, 218
Laplace Transform, 181, 183, 185, 186,
    188, 189, 193, 204
Lead Time, 225
Learning Effect, 225
Learning Rate, 228
Lime, 137
Lot Size, 225

Markov Chain, 1, 3
Measurement Errors, 1, 2, 3, 4, 5, 7, 11,
    12, 14, 15, 16, 19

Nadaraya-Watson Estimator, 3,
    10, 19
Naïve Estimator, 91